Christian Cremer

**Neurotransmitter Receptors in Animal Models of Neurodegeneration**

Christian Cremer

# Neurotransmitter Receptors in Animal Models of Neurodegeneration

## Anatomy, Pharmacology and Molecular Properties

Südwestdeutscher Verlag für Hochschulschriften

**Impressum/Imprint (nur für Deutschland/only for Germany)**
Bibliografische Information der Deutschen Nationalbibliothek: Die Deutsche Nationalbibliothek verzeichnet diese Publikation in der Deutschen Nationalbibliografie; detaillierte bibliografische Daten sind im Internet über http://dnb.d-nb.de abrufbar.
Alle in diesem Buch genannten Marken und Produktnamen unterliegen warenzeichen-, marken- oder patentrechtlichem Schutz bzw. sind Warenzeichen oder eingetragene Warenzeichen der jeweiligen Inhaber. Die Wiedergabe von Marken, Produktnamen, Gebrauchsnamen, Handelsnamen, Warenbezeichnungen u.s.w. in diesem Werk berechtigt auch ohne besondere Kennzeichnung nicht zu der Annahme, dass solche Namen im Sinne der Warenzeichen- und Markenschutzgesetzgebung als frei zu betrachten wären und daher von jedermann benutzt werden dürften.

Coverbild: www.ingimage.com

Verlag: Südwestdeutscher Verlag für Hochschulschriften GmbH & Co. KG
Dudweiler Landstr. 99, 66123 Saarbrücken, Deutschland
Telefon +49 681 37 20 271-1, Telefax +49 681 37 20 271-0
Email: info@svh-verlag.de

Approved by: Düsseldorf, HHU, Diss., 2010

Herstellung in Deutschland:
Schaltungsdienst Lange o.H.G., Berlin
Books on Demand GmbH, Norderstedt
Reha GmbH, Saarbrücken
Amazon Distribution GmbH, Leipzig
**ISBN: 978-3-8381-2831-3**

**Imprint (only for USA, GB)**
Bibliographic information published by the Deutsche Nationalbibliothek: The Deutsche Nationalbibliothek lists this publication in the Deutsche Nationalbibliografie; detailed bibliographic data are available in the Internet at http://dnb.d-nb.de.
Any brand names and product names mentioned in this book are subject to trademark, brand or patent protection and are trademarks or registered trademarks of their respective holders. The use of brand names, product names, common names, trade names, product descriptions etc. even without a particular marking in this works is in no way to be construed to mean that such names may be regarded as unrestricted in respect of trademark and brand protection legislation and could thus be used by anyone.

Cover image: www.ingimage.com

Publisher: Südwestdeutscher Verlag für Hochschulschriften GmbH & Co. KG
Dudweiler Landstr. 99, 66123 Saarbrücken, Germany
Phone +49 681 37 20 271-1, Fax +49 681 37 20 271-0
Email: info@svh-verlag.de

Printed in the U.S.A.
Printed in the U.K. by (see last page)
**ISBN: 978-3-8381-2831-3**

Copyright © 2011 by the author and Südwestdeutscher Verlag für Hochschulschriften GmbH & Co. KG and licensors
All rights reserved. Saarbrücken 2011

**Abstract**

Neurodegeneration comprehends all processes of progressive loss of neuronal structure or function in the vertebrate nervous system. As a common hallmark, neurodegenerative processes relate otherwise dissimilar disorders like epilepsy, Parkinson's or Alzheimer's disease. Neurotransmitter receptors are key elements of synaptic transmission that link neuronal structure and function. Thus, a comprehensive analysis of the distribution pattern, regional densities and pharmacological properties of neurotransmitter receptors in animal models of neurodegeneration could reveal underlying molecular mechanisms. Here, changes of neurotransmitter receptors were studied in rodent models of repeated seizures, disturbed neurotransmitter homeostasis or reelin gene mutation, respectively.

The convulsant pentylenetetrazole (PTZ) was used to model epileptic seizures in rats. Alterations of neurotransmitter receptor densities were quantified (Cremer et al., 2009a). A general reduction of kainate receptors was observed together with a regional specific increase of NMDA and $GABA_A$ associated benzodiazepine (BZ) binding sites and decreased adenosine $A_1$ receptor binding.

According to previous studies, the astrocytic enzyme glutamine synthetase (GS) becomes nitrated and partially inhibited in the PTZ seizure model. GS is a key regulator of glutamate and GABA metabolism in the glutamate/glutamine cycle. Since changes of neurotransmitter receptor densities were demonstrated in the PTZ model, similar changes were hypothesized when GS is inhibited in vivo. Therefore, rats were treated with L-methionine sulfoximine (MSO), an irreversible inhibitor of GS. Changes of neurotransmitter receptor densities and subunit expression were studied (Cremer et al., 2010a). A significant, regional specific reduction of BZ binding was found and concomitant, but differential changes of $GABA_A$ subunit composition.

As a prerequisite to quantify receptor subunit mRNA, a novel quantitative in situ hybridization (ISH) protocol was established. To evaluate this method, mRNAs of the AMPA receptor subunits GluR1 and GluR2 were measured in rats treated with the organo-arsenic compound dimethyl-arsenic acid ($DMA^{III}$), which is known to reduce the number of AMPA receptors in the brain. Accordingly, significant reductions of GluR1 and GluR2 subunit expression in the hippocampus of DMA-treated rats were found (Cremer et al., 2009b).

Neurodegeneration may result from genetic aberrations leading to changes of neuronal structure or function. In mice, a mutation of the extracellular matrix protein *reelin* leads to developmental deficits of neuronal migration causing cerebellar hypoplasia, disturbed laminar pattern of the hippocampus and an inversion of neocortical layers, resulting in the so called "reeler" phenotype. In the adult brain, *reelin* regulates synaptic plasticity by modulating neurotransmitter receptor function. Thus, the effects of reelin mutation on neurotransmitter receptor densities and distribution in reeler mice were studied (Cremer et al., 2010b). Differential changes were demonstrated in the laminar distribution, maximum binding capacity ($B_{max}$) and regional density of several neurotransmitter receptors in the reeler brain, indicating a role for *reelin* in neurotransmitter receptor expression.

Taken together, the investigation of neurotransmitter receptors in different animal models of neurodegeneration demonstrated that i) loss of neuronal structure or function coincided with differential changes of neurotransmitter receptor densities in all investigated models ii) correlating changes of receptor densities could occur in numerous brain regions or in a regionally restricted manner iii) a disturbed neuronal function could influence receptor subunit composition and mRNA expression. Conclusively, this study revealed a complex pattern of correlations between neurodegeneration and changes of neurotransmitter receptors. Since neurotransmitter receptors are a major target for pharmacological intervention, these results might offer ambitions for the development of therapeutic strategies.

**Zusammenfassung**

Neurodegeneration vereint als Sammelbegriff alle Vorgänge progressiven Verlusts neuronaler Struktur oder Funktion im Nervensystem von Vertebraten. Diese Prozesse sind gemeinsames Merkmal ansonsten unterschiedlicher Erkrankungen wie Epilepsie, Morbus Parkinson oder Morbus Alzheimer. Neurotransmitterrezeptoren sind Schlüsselelemente synaptischer Übertragung und werden daher als zentrales Bindeglied zwischen neuronaler Struktur und Funktion betrachtet. Eine umfangreiche Analyse der Verteilungsmuster, regionalen Dichten und pharmakologischen Eigenschaften von Neurotransmitterrezeptoren in verschiedenen Modellen von Neurodegeneration könnte daher zugrundeliegende Mechanismen offenbaren. In dieser Arbeit wurden Veränderungen von Neurotransmitterrezeptoren in Nagermodellen der Epilepsie, gestörter Neurotransmitterhomöostase oder Reelin-Mutation studiert.

Als Modell für Epilepsie wurden Ratten mit dem Konvulsivum Pentylentetrazol (PTZ) behandelt. Die damit einhergehenden Veränderungen der Neurotransmitterrezeptordichten wurden untersucht (Cremer et al., 2009a). Hierbei wurde eine generelle Verringerung der Kainat-Rezeptordichten im Gehirn der behandelten Ratten festgestellt. Diese korrelierte mit regionenspezifischen Dichteerhöhungen der NMDA- und $GABA_A$ assoziierten Benzodiazepinbindestellen bzw. mit Verringerungen der Adenosin $A_1$ Rezeptordichte.

Frühere Studien zeigten, dass im PTZ-Modell das astrozytenspezifische Enzym Glutaminsynthetase (GS) verstärkt nitriert und funktionell inhibiert ist. Die GS ist ein Schlüsselenzym des Glutamat- und GABA-Metabolismus im Glutamat/Glutamin-Zyklus. Da Veränderungen der Neurotransmitterrezeptordichten im PTZ Modell beobachtet wurden, könnte eine Inhibition der GS zu ähnlichen Veränderungen führen. Es wurden daher Ratten mit L-Methinin-Sulfoximin behandelt, einem irreversiblen Inhibitor der GS, und die daraus resultierenden Veränderungen der Rezeptordichte und deren Untereinheiten untersucht (Cremer et al., 2010a). Eine Inhibition der GS führte zu einer signifikanten, regionenspezifischen Verringerung der $GABA_A$ assoziierten Benzodiazepinbindestellen und einhergehenden, differentiellen Veränderungen der $GABA_A$ Untereinheitenzusammensetzung.

Um die mRNA-Expression von Rezeptoruntereinheiten quantifizieren zu können, wurde ein optimiertes Protokoll der quantitativen in situ Hybridisierung etabliert. Zur

Evaluation dieser Methode wurden die mRNA Level der Glutamatrezeptoruntereinheiten GluR1 und GluR2 im Gehirn von Ratten bestimmt, die einer Behandlung mit Dimethyl-Arsensäure unterzogen wurden, welche bekanntermaßen die regionale Dichte von AMPA-Rezeptoren verringert. Auf mRNA Ebene wurde eine entsprechende Verringerungen der GluR1 und GluR2 Untereinheitenexpression im Hippocampus DMA$^{III}$–behandelter Ratten festgestellt (Cremer et al., 2009b).

Neurodegeneration kann Folge genetischer Aberrationen sein und somit zu Veränderungen neuronaler Struktur und Funktion führen. Eine Mutation des extrazellulären Matrixproteins Reelin führt in sog. Reeler-Mäusen zu Defiziten der neuronalen Migration während der Ontogenese, und in der Folge zu zerebellärer Hypoplasie, gestörter Laminierung des Hippocampus und einer invertierten Orientierung der neokortikalen Schichten. Im adulten Gehirn ist Reelin an der Regulation synaptischer Plastizität durch Neurotransmitterrezeptormodulation beteiligt. Daher wurden im Rahmen dieser Arbeit die Neurotransmitterrezeptordichten und ihre Verteilung in Reeler-Mäusen untersucht (Cremer et al., 2010b). Hierbei wurden komplexe Veränderungen der laminären Rezeptorverteilung, der maximalen Bindekapazität ($B_{max}$), und der regionenspezifischen Rezeptordichte gemessen, die eine Beteiligung Reelins bei der Expression von Neurotransmitterrezeptoren implizieren.

Die Untersuchung von Neurotransmitterrezeptoren in Tiermodellen der Neurodegenration zeigte, i) dass der Verlust neuronaler Struktur oder Funktion in allen Modellen mit Veränderungen der Neurotransmitterrezeptordichten einherging ii) dass korrelierende Veränderungen von Rezeptordichten sowohl regional begrenzt als auch über verschiedene Areale hinweg auftreten konnten iii) dass eine gestörte neuronale Funktion zu Veränderungen der Untereinheitenzusammensetzung und mRNA-Expression von Neurotransmitterrezeptoren führte. Zusammengefasst zeigte diese Studie ein komplexes Korrelationsmuster zwischen Neurodegeration und Veränderungen von Neurotransmitterrezeptoren. Da Neurotransmitterrezeptoren ein zentrales Ziel pharmakologischer Intervention darstellen, offerieren diese Ergebnisse mögliche Perspektiven für neuartige Therapieansätze neurodegenerativer Erkrankungen.

# Neurotransmitter Receptors in Animal Models of Neurodegeneration: Anatomy, Pharmacology and Molecular Properties

**Preface: Organization of the manuscript**

**Introduction**

    1. Framework      1

    2. Neurotransmitter receptors      -

    3. Animal models

        3.1 Pentylenetetrazole-induced seizures      3

        3.2 Inhibition of glutamine synthetase      4

        3.3 Reeler mice      7

    4. Quantitative in situ hybridization      11

    5. Aims of the study      13

**Publications**

    Cremer et al., 2009a      15
    *Pentylenetetrazole-induced seizures affect binding site densities for GABA, glutamate and adenosine receptors in the rat brain.*

    Cremer et al., 2009b      43
    *Fast, quantitative in situ hybridization of rare mRNAs using $^{14}C$-standards and phosphorus imaging.*

    Cremer et al., 2010a      65
    *Inhibition of glutamate/glutamine cycle in vivo results in decreased benzodiazepine binding and differentially regulated GABAergic subunit expression in the rat brain.*

    Cremer et al., 2010b      99
    *Laminar distribution of neurotransmitter receptors in the reeler mouse cerebral cortex*

**Discussion**

    1. Pentylenetetrazole-induced seizures      139

Table of contents

|   |   |
|---|---|
| 2. Inhibition of glutamine synthetase | 141 |
| 3. Reeler mice | 143 |
| 4. Quantitative in situ hybridization | 146 |
| 5. Summary and conclusions | 147 |

**References** 151

**Acknowledgements** 163

**Addendum** 165

Cremer CM, Palomero-Gallagher N, Bidmon HJ, Schleicher A, Speckmann EJ, Zilles K (2009) Pentylenetetrazole-induced seizures affect binding site densities for GABA, glutamate and adenosine receptors in the rat brain. Neuroscience 163(1):490-9.

Cremer CM, Cremer M, Escobar JL, Speckmann EJ, Zilles K (2009) Fast, quantitative in situ hybridization of rare mRNAs using 14C-standards and phosphorus imaging. J Neurosci Methods 185:56-61.

Cremer CM, Bidmon HJ, Görg B, Palomero-Gallagher N, Escobar JL,Speckmann EJ, Zilles K (2010) Inhibition of glutamate/glutamine cycle in vivo results in decreased benzodiazepine binding and differentially regulated GABAergic subunit expression in the rat brain. Epilepsia, in press [uncorrected proof].

**Preface: Organization of the manuscript**

The present work is a cumulative dissertation, based on peer-reviewed manuscripts submitted to different international journals in the field of neuroscience.

A general introduction integrates the manuscripts into a common scientific context. Additionally, the goals of the present work are defined by common theses. All articles are briefly introduced separately with emphasis on the underlying hypothesis and the essential conclusions. The author's contribution to the work is addressed in each case as well as the current state of publication (i.e. in review, in press, published).

The articles are provided in chronological order in the author's manuscript version as submitted/accepted by the respective journal. Thus, the present paper exhibits a uniform appearance independent of editing by the respective journals. However, all contents (text, figures and tables) of each manuscript are *de facto* identical to the edited journal version. Articles published or in press are additionally provided in the edited version in the addendum.

# Introduction

## 1. Framework

Neurodegeneration is a comprehensive term for the progressive loss of neuronal structure or function. Among others, neurodegenerative processes relate otherwise dissimilar disorders like epilepsy, Parkinson's or Alzheimer's disease. Neurotransmitter receptors mediate synaptic transmission and therefore act as a link between neuronal organization and signal transduction in the brain. Thus, in the present work neurodegeneration particularly refers to processes which affect the expression pattern, regional densities or pharmacological properties of neurotransmitter receptors.

The study of animal models which exhibit disturbances of neuronal structure or function helps to reveal underlying molecular mechanisms. This work presents a comprehensive analysis of neurotransmitter receptors in rodent models of repeated seizures, disturbed neurotransmitter homeostasis or reelin gene mutation, respectively. Additionally, a new method of quantitative in situ hybridization was established, which allows a more differential study of receptor subunit expression.

## 2. Neurotransmitter receptors

Neurotransmitter receptors are key elements of synaptic transmission that mediate signal transduction by binding of presynaptically released neurotransmitters. Furthermore, the initiation of synaptic plasticity, a molecular correlate of learning and memory formation, decisively depends on the involvement of neurotransmitter receptors (Kandel et al., 2000). Local changes in the density and distribution of receptors coincide with cytoarchitectonically defined areas (reviewed by Zilles and Amunts, 2009a, b; Zilles et al., 2004). Therefore, their analysis has become an important pillar of modern neuroanatomy.

Due to their relevance for proper neuronal function, changes of neurotransmitter receptors were demonstrated to correlate with neurological disorders or respective experimental conditions in rodents. For example, in epileptic brain tissue differential alterations of neurotransmitter receptor systems were found. Neocortical tissue explants from patients suffering from focal epilepsy exhibited an increase of α-amino-3-hydroxy-5-methyl-4-isoxazole propionic acid (AMPA) binding sites together with non-uniform alterations of γ-amino butyric acid (GABA) receptors (Zilles et al.,

1999). Furthermore, an increase of AMPA-, kainate and N-methyl-D-aspartate (NMDA) receptors was shown in temporal lobe epilepsy, together with increased $GABA_B$ receptor densities (Bidmon et al., 2002). These changes were accompanied by enhanced heat shock protein expression in the neocortex (Bidmon et a., 2004). These observations were later confirmed in an animal model of epilepsy (Bidmon et al., 2005) and were shown to correlate with reduced glutamine synthetase activity (Bidmon et al., 2008).

The results noted above particularly demonstrate that changes of a single receptor system most often result in accompanying alterations of other systems. Thus, a comprehensive approach is needed to reliably study neurotransmitter receptor expression. Quantitative *in vitro* autoradiography is a powerful tool to visualize and quantify neurotransmitter receptors in native brain sections, allowing a comprehensive analysis of their density and distribution pattern. In the present work quantitative receptor autoradiography was used together with different pharmacological, immunohistochemical and biomolecular methods to study different rodent models of disturbed neuronal structure or function. In particular, the effects of repeated seizures, disturbed neurotransmitter homeostasis or reelin gene mutation were analyzed.

An imbalance between neuronal excitation and inhibition may cause epileptic seizures. Repeated intraperitoneal injection of the convulsant pentylenetetrazole (PTZ), an inhibitor of $GABA_A$ receptors, results in acute and chronic seizures in rats. Thus, PTZ treated rats were used as a model to study seizure related changes of neurotransmitter receptors.

The orchestrated release, reuptake and recycling of neurotransmitters are prerequisites for proper synaptic transmission. Disruption of this chain of events might lead to disturbed neurotransmitter homeostasis and eventually changes of receptor expression. The astrocytic enzyme glutamine synthetase (GS) is a key element of neurotransmitter recycling in the glutamate/glutamine cycle. Thus, *in vivo* inhibition of GS was used to study the effect of disturbed glutamate and GABA recycling on neurotransmitter receptor expression.

The reeler mouse was used as a reference model of disturbed neuronal organization. In these mice, a lack of the extracellular matrix protein reelin leads to deficits of neuronal migration during development, resulting in severe changes of cyto- and myeloarchitecture (reviewed by de Rouvroit and Goffinet, 1998). Additionally, reelin enhances synaptic efficacy by modulation of neurotransmitter receptor function (reviewed by Herz and Chen, 2006). Thus, reelin deficiency may be associated with alterations of neurotransmitter receptor densities and distribution.

## 3. Animal models

In the present work, different animal models were studied. In the following, the basic properties of the respective models will be introduced. All experiments were performed according to the German animal welfare act and the guidelines for the treatment of experimental animals of the Research Center Jülich and the Heinrich-Heine University, Düsseldorf. All animal experiments were approved by the responsible governmental agency.

### 3.1 Pentylenetetrazole-induced seizures

Intraperitoneal applications of PTZ are used to investigate the effects of acute and chronic epileptic seizures in rodents (Caspers and Speckmann, 1972; Bertram, 2007). PTZ is a $GABA_A$ receptor antagonist (Macdonald and Barker, 1978) which binds at benzodiazepine (BZ) binding sites of the functional $GABA_A$ receptor (Rehavi et al., 1982; Squires et al., 1984; Huang et al. 2001). Binding sites for GABA and BZs have different localization and structural properties at the functional $GABA_A$ receptor: The BZ binding site is located between $\gamma_2$ and $\alpha_{(1,2,3,5)}$ subunits of the pentameric $GABA_A$ receptor, while GABA (as well as the [$^3$H]-ligand muscimol) binds between $\alpha$ and $\beta$ subunits of the receptor (Pritchett et al., 1989; Rudolph et al., 1999; McKernan et al., 2000; Baumann et al., 2003). Binding of BZ is known to enhance $GABA_A$ receptor activity (Figure 1).

PTZ treatment in mice and rats has numerously been shown to exhibit alterations of glutamatergic, GABAergic and adenosinergic receptors (Angelatou, 1990; Pagonopoulou et al., 1993; Ekonomou and Angelatou, 1999; Walsh et al., 1999; Ekonomou et al., 2001; Tchekalarova, 2005). An upregulation of AMPA binding was shown (Ekonomou et al., 2001), which depends on the duration of the PTZ-treatment, as well as a long lasting upregulation of NMDA binding sites (Ekonomou and Angelatou, 1999). The $GABA_A$ receptor subunit composition was transiently affected with minimal or no effect on $GABA_A$ receptor or BZ binding site densities (Walsh et al., 1999). A regionally restricted long-term increase of adenosine type 1

($A_1$) receptor binding was also reported (Angelatou et al., 1990; Pagonopoulou et al., 1993; Tchekalarova et al., 2005), while a persistent down-regulation of $A_1$ receptors was found in the basal ganglia (Pagonopoulou et al., 1993). However, most of the studies noted above emphasize only few receptor types or brain regions, respectively. Thus, we comprehensively investigated regional binding site densities of AMPA, kainate, NMDA, $A_1$, $GABA_A$, and $GABA_B$ receptors as well as of BZ binding sites in brains of PTZ-treated rats.

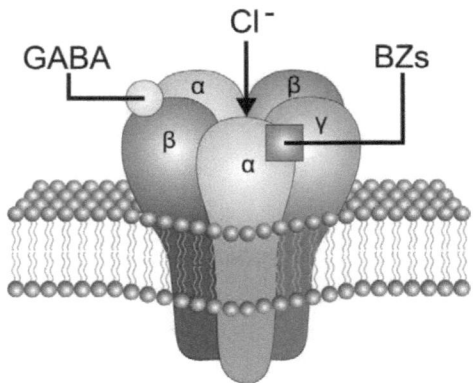

**Figure 1:** $GABA_A$ receptors belong to the Cys–loop family of transmitter-gated ion channels and contain five membrane-crossing subunits that are arranged to form an intrinsic anion-conducting channel. There are 19 subunits (α1–6, β1–3, γ1–3, δ, ε, θ, π and ρ1–3), which are classified according to sequence homology. Most $GABA_A$ receptors *in vivo* comprise α-, β- and γ-subunits in a probable stoichiometry of 2α:2β:1γ. The most prevalent subtype is the $α_1β_2γ_2$ isoform (~60%). The subunit composition influences fundamental features of the receptor, including sensitivity to GABA, channel kinetics and desensitization, neuronal location and pharmacological properties. Based on Belelli and Lambert, 2005.

## 3.2 Inhibition of glutamine synthetase

MSO irreversibly inhibits the astrocytic enzyme glutamine synthetase, a key regulator of glutamate and GABA metabolism in the glutamate/glutamine cycle (GGC) (Figure 2). It has been demonstrated that disturbances of the GGC result in differential changes of neurotransmitter homeostasis. Application of MSO *in vivo* transiently decreases whole brain GABA content (Stransky, 1969). GABA, glutamate and glutamine concentrations are differentially altered in the neostriatum and globus pallidus of MSO-treated rats (Fonnum and Paulsen, 1990), while the striatal release

Introduction

of dopamine is affected by MSO infusion (Rothstein and Tabakoff, 1982). Furthermore, the GGC regulates synaptic GABA content under physiological conditions (Liang et al., 2006).

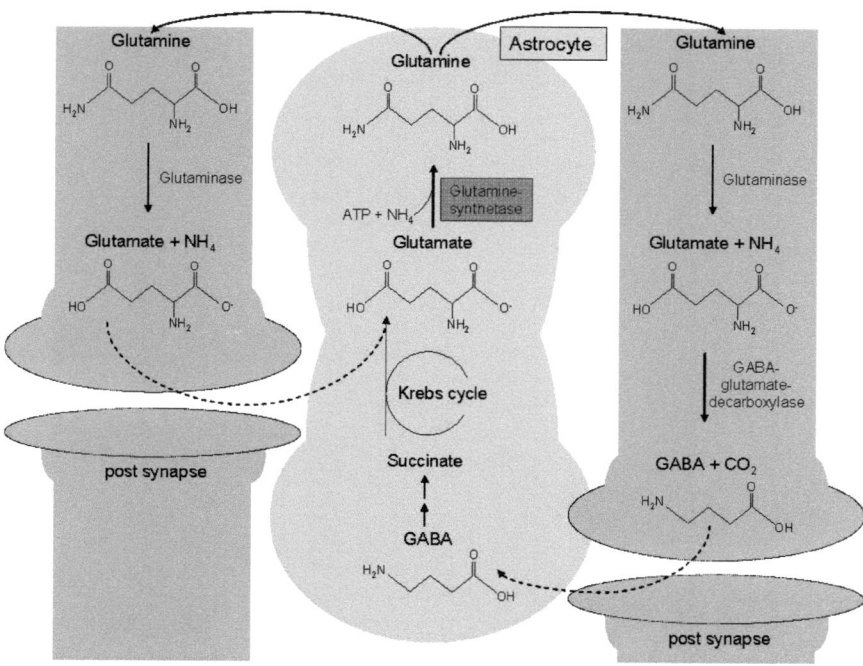

**Figure 2:** Schematic summary of neurotransmitter recycling in the glutamate/glutamine cycle. Glutamate is released into the synaptic cleft during neurotransmission and subsequently removed via uptake into astrocytes. The astrocytic enzyme glutamine synthetase catalyzes the generation of glutamine from glutamate (Martinez-Hernandez et al., 1977), which is transferred via system N and A transporters into neurons (Chaudhry et al., 2002) where it is deaminated by phosphate-activated glutaminase, thus producing glutamate (Kvamme et al., 2000). Although GABAergic interneurons are capable of recycling GABA by specific transporters, a major source of vesicular GABA is derived from decarboxylation of glutamate (Martin and Tobin, 2000). Methionine sulfoximine irreversibly inhibits the function of glutamine synthetase *in vivo* and *in vitro* (Lamar and Sellinger, 1965; Ronzio et al., 1969).

Previous reports indicated that in the PTZ model of epilepsy GS is nitrated and partially inhibited (Bidmon et al., 2005) and we demonstrated alterations of neurotransmitter receptor expression in the same model (Cremer et al., 2009a). Together, these evidences led to the hypothesis that inhibition of MSO *in vivo* might result in changes of neurotransmitter receptor densities. We used $^3$H-receptor autoradiography to measure glutamate (AMPA, kainate, NMDA), GABA ($GABA_A$, $GABA_B$ and BZ binding sites), dopamine ($D_1$) and adenosine ($A_1$) receptor subtypes in different brain regions of MSO treated rats. Additionally, we performed saturation analysis of BZ binding sites on cerebral membrane homogenates and investigated the expression of $GABA_A$ $\alpha_1$ and $\gamma_2$ subunits, which primarily mediate BZ binding, by western blot analysis and quantitative in situ hybridization.

## 3.3 Reeler mice

Reelin is a major secretory glycoprotein with important roles in embryogenesis and during adulthood (reviewed by de Rouvroit and Goffinet, 1998; Levenson et al., 2008). Mutation of reelin leads to severe neurodevelopmental deficits resulting in cerebellar hypoplasia, mislamination of the hippocampus and an inversion of neocortical layers (Figures 3 and 4). In the adult brain reelin plays a significant role in cellular maturation, synaptic plasticity and learning (reviewed by Herz and Chen, 2006; Rogers and Weeber, 2008).

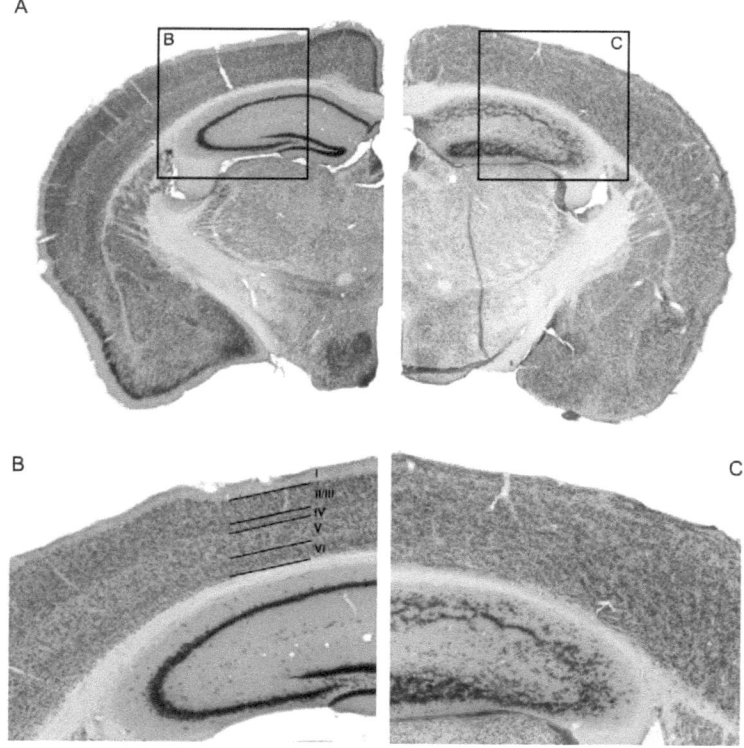

**Figure 3:** Frontal sections of wild type (left) and Reeler brains (right) stained immunohistochemically for neuronal nuclei (NeuN). A) Frontal sections (Bregma -1.8 mm) through whole hemispheres demonstrate striking differences in cytoarchitecture between wild type and Reeler brains. B) Magnified area as indicated in A. from an adjacent section. Cortical lamination is plotted as revealed by microscopic examination. C) Magnified area as indicated above. Cortical lamination is severely disturbed in Reeler. Modified from Cremer, 2007.

Introduction

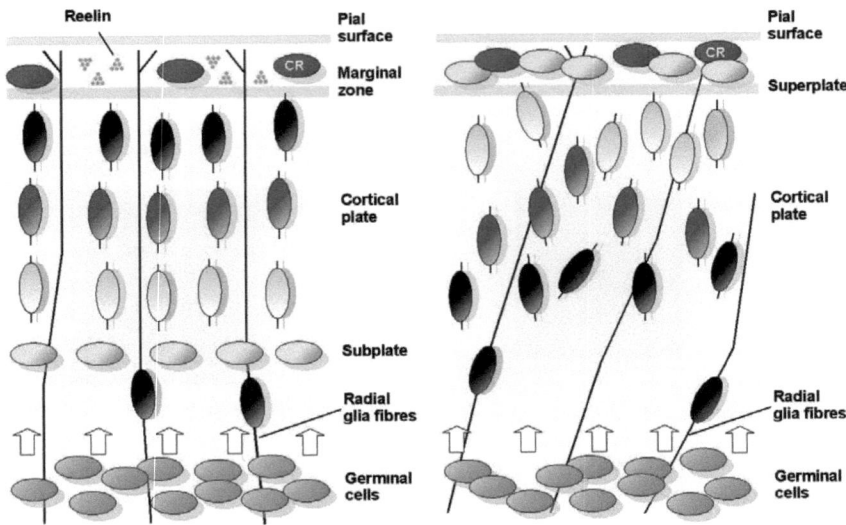

**Figure 4:** Schematical summary of cortical development in wild type (left) and Reeler mice (right). Cortical neurons are generated in the germinal zone and migrate radially towards the pial surface along radial glia fibers, thereby splitting the preplate into a subplate and a marginal zone. Originating neurons migrate through earlier generated layers and detach from radial glia fibers directly below Cajal-Retzius (CR) cell layer (later layer I), resulting in an "inside first, outside last" generation of cortical layers. In Reeler mice neurons from the germinal cell layer are unable to split the malformed preplate or superplate and often do not detach properly from the obliquely oriented radial glia fibers. Originating neurons do not migrate through earlier layers but reside directly below, resulting in an "outside first, inside last" developmental pattern. Modified from Cremer, 2007.

In recent years research on reelin has come to a crossroad. On the one hand the reeler mouse has been proposed as a model to study certain aspects of different neuronal diseases like lissencephaly, epilepsy, schizophrenia and Alzheimer's disease (reviewed by Fatemi et al., 2008). For example, only recently it was demonstrated that reduced reelin expression accelerates amyloid-ß-plaque formation and tau pathology in transgenic Alzheimer's disease (AD) mice (Kocherhans et al., 2010). Furthermore, reelin signaling antagonizes beta-amyloid-induced suppression of NMDA receptor-mediated long-term potentiation (Durakoglugil et al., 2009).

On the other hand effort has been made to shed light on the molecular pathway underlying reelin function in the adult brain (reviewed by Rogers and Weeber, 2008; Herz and Chen, 2006), with emphasis on regulation of neurotransmitter receptor activity (Figure 5). Binding of reelin to the lipoprotein receptors Apolipoprotein E Receptor 2 (apoER2) and/or Very Low Density Lipoprotein Receptor (VLDLR) results in enhanced NMDA receptor activity, mediated by subunit phosphorylation through a Src kinase. Activation of the reelin pathway results in both increased conductance and increased trafficking of AMPA receptors to the membrane. Additionally, reelin regulates microtubule organization by downstream activation of cyclin-dependent kinase 5 (CDK-5) and inhibition of tau phosphorylation.

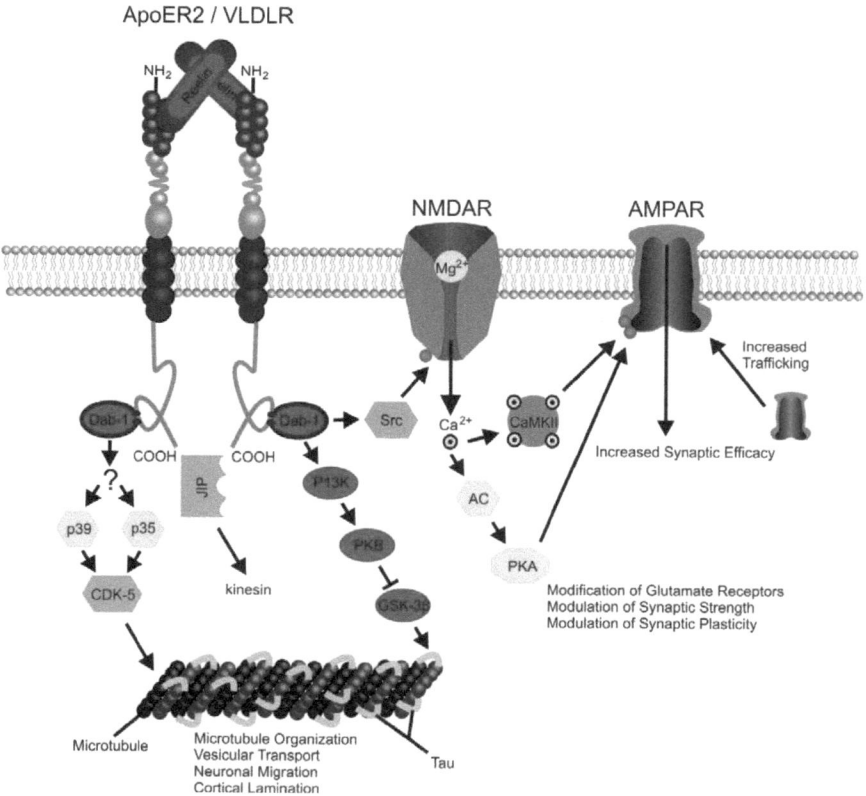

**Figure 5**: Model for reelin-mediated enhancement of synaptic efficacy. Reelin binds to its constituent receptors Apolipoprotein E Receptor 2 (apoER2) and/or Very Low Density Lipoprotein Receptor (VLDLR) as a dimer. Both reelin receptors are coupled to intracellular signaling machinery via the protein Disabled-1 (Dab1). Once Dab1 is activated, a host of signaling commences through Src, PI3K and CDK5. Src phosphorylates NMDA receptor subunits on tyrosine residues, resulting in enhancement of NMDA receptor function. This pathway is responsible for the reelin-induced

## Introduction

enhancement of induction of NMDA receptor dependent synaptic plasticity. Downstream of the NMDA receptor, CaMKII and PKA phosphorylate the AMPA receptor resulting in both increased conductance and increased trafficking of these receptors to the membrane. Based on Levenson et al., 2008.

The implications of reelin function for receptor regulation as well as a disturbed neuronal organization in reeler mutant mice led to the hypothesis that reelin deficiency may be associated with alterations of neurotransmitter receptor densities or pharmacological properties. Thus, we conducted a multi-receptor study to investigate the influence of the *reelin* mutation on various neurotransmitter receptors in the adult brain by means of *in vitro* receptor autoradiography. Furthermore, saturation analysis were performed for each receptor type to determine receptor binding affinity (dissociation constant, $K_D$) and maximum binding capacity ($B_{max}$).

## 4. Quantitative in situ hybridization

Quantitative receptor autoradiography is a pharmacological approach to reliably investigate the density and distribution of neurotransmitter receptors in the brain. A major advantage of this method is the fact that only functional receptors residing at the neuronal membrane are measured, while internalized (e.g. trafficking) or intracellularly stored receptors or its precursors remain disregarded.

However, this method does not allow conclusions on changes of receptor subunit composition or mRNA transcription level. A combination of classical receptor autoradiography and quantification of the respective subunit mRNAs would allow a more differential study and shed light on the molecular mechanisms of receptor density alterations.

Quantitative in situ hybridization (qISH) of mRNAs using radiolabeled complementary nucleic acid probes is a powerful method to visualize and quantify gene expression. The measurement of mRNAs using $^{33}$P- or $^{35}$S-labelled oligodeoxyribonucleotides is a highly sensitive, reliable technique and bears the advantage of visualizing mRNA expression within the native tissue enabling a discrete anatomical correlation. Although qISH is widely accepted as an appropriate method to analyze gene expression patterns in the brain, not all current protocols are applicable concerning standardization, sensitivity and signal to noise ratio. However, protocols matching all premises for accurate qISH are often laborious and time demanding.

To overcome these disadvantages, we established a method for fast qISH of mRNAs using $^{33}$P-labeled oligodeoxyribonucleotides together with $^{14}$C-polymer standards and a phosphorus imaging system (Cremer et al., 2009b). Our approach can easily be performed in any laboratory to quantify the expression of high and low abundant mRNAs and reduces the time needed for qISH from several weeks to few days.

To test our method, we used hippocampal sections of rats, treated with the organo-arsenic compound dimethyl arsenic acid (DMA$^{III}$). DMA$^{III}$ is known to drastically reduce the number of AMPA receptors in the rat hippocampus (Lopez-Escobar, 2009; Figure 6). These findings have earlier been suggested by electrophysiological observations, e.g. excitatory postsynaptic potentials were reduced in amplitude or completely inhibited in hippocampal slice culture after application of different organo-arsenic compounds (Krüger et al., 2006; 2007). We hypothesized that the observed decrease of AMPA receptor density might be accompanied by changes of the receptor subunit mRNA expression.

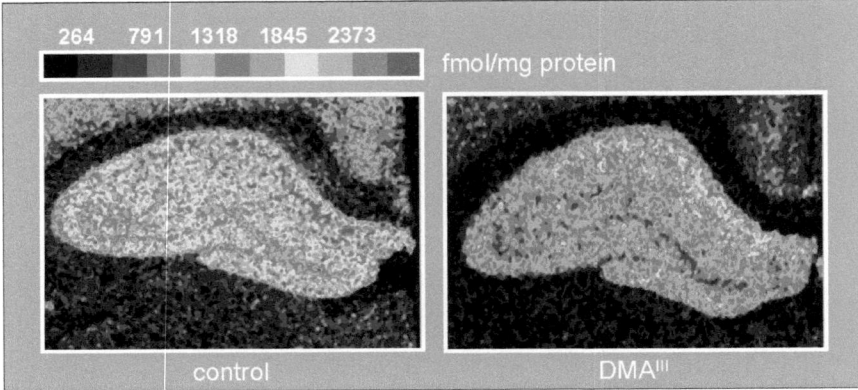

**Figure 6:** Color coded autoradiographs of AMPA receptor densities in the hippocampus of male, adult rats treated at postnatal day 7 with either a single intraperitoneal injection of the organo-arsenic compound DMA$^{III}$ or saline as a control. DMA$^{III}$ treatment results in a drastic reduction (25-28% decrease) of AMPA receptor densities (labeled with $^3$H-AMPA) in all investigated regions. Modified from Lopez Escobar, 2009.

**Aims of the study**

In the present work different animal models were used to investigate the influence of repeated seizures, disturbed neurotransmitter homeostasis or reelin gene mutation on neurotransmitter receptors in the brain. Several technical approaches were used to study the density, distribution, pharmacological properties and subunit expression of neurotransmitter receptors under different experimental conditions. Furthermore, a reliable method was established to quantify the mRNA expression of neurotransmitter receptors or subunits, respectively.

While each study had a defined focus resulting in specific hypotheses and methodological approaches, the following aims and questions were underlying the analysis of all studies.

1. Are the densities and regional distribution of neurotransmitter receptors affected in a given animal model?

2. Are the observed changes restricted to a single receptor type or a specific brain region?

3. Do changes of receptor density correlate with altered receptor pharmacology (e.g. binding site affinity, maximum binding capacity)?

4. Do changes of receptor subunit composition or mRNA transcription levels play a role in receptor regulation?

## Pentylenetetrazole-induced seizures affect binding site densities for GABA, glutamate and adenosine receptors in the rat brain

Christian M. Cremer; Nicola Palomero-Gallagher; Hans-Jürgen Bidmon; Axel Schleicher; Erwin-J. Speckmann and Karl Zilles

Pentylenetetrazole (PTZ) is a convulsant used to model epileptic seizures in rats. In the PTZ-model, altered heat shock protein 27 (HSP-27) expression highlights seizure affected astrocytes, which play an important role in glutamate and GABA metabolism. This raises the question whether impaired neurotransmitter metabolism leads to an imbalance in neurotransmitter receptor expression. Consequently, we investigated the effects of seizures on the densities of seven different neurotransmitter receptors in rats which were repeatedly treated with PTZ (40 mg/kg) over a period of 14 days.

Our data demonstrate the impact of PTZ induced seizures on the densities of kainate, NMDA, $A_1$ and BZ binding sites in epileptic brain. These changes are not restricted to regions showing glial impairment.

Contributions on experimental design, realization and publication:
Histological and autoradiographic data acquisition and analysis were accomplished by C. Cremer. Autoradiographic and histological stainings were contributed by technicians M. Cremer, S. Krause and S. Wilms who are acknowledged in the manuscript. The manuscript including all figures and tables was prepared by C. Cremer and subsequently reviewed, amended and approved by all co-authors.

    Approximated total share of contribution in per cent: 60%

    Current state of publication: published. Neuroscience 2009; 163(1):490-9.

Publications

## Pentylenetetrazole-induced seizures affect binding site densities for GABA, glutamate and adenosine receptors in the rat brain

*Christian M. Cremer[a,b]; Nicola Palomero-Gallagher[a]; Hans-Jürgen Bidmon[b]; Axel Schleicher[b]; Erwin-J. Speckmann[c] and Karl Zilles[a,b,d]*

[a]*Institute for Neuroscience and Medicine (INM-2), Research Center Jülich, D-52425 Jülich, Germany*

[b]C. & O. Institute for Brain Research, Heinrich-Heine-University Düsseldorf, Universitätsstr. 1, Bldg. 22.03.05, D-40225 Düsseldorf, Germany

[c]Institute of Physiology I, Westfälische-Wilhelms-University, Robert-Koch-Str. 27a, D-48149 Münster, Germany

[d]JARA, Jülich Aachen Research Alliance, Research Center Jülich, D-52425 Jülich, Germany

Corresponding author:

*Christian Cremer*

*Institute for Neuroscience and Medicine (INM-2)*

*Research Center Jülich, D-52425 Jülich, Germany*
Phone: +49-2461-61-8615 Fax: +49-2461-61-2820
Email: c.cremer@fz-juelich.de

Section Editor:
Dr. Asla Pitkänen
Department of Neurobiology
A.I. Virtanen Institute for Molecular Sciences
University of Kuopio, PO Box 1627, FIN-70211 Kuopio, Finland

## Abbreviations

| | |
|---|---|
| $A_1$ | adenosine type 1 (receptor) |
| AMG | amygdala |
| AMPA | α-amino-3-hydroxy-5-methyl-4-isoxazole propionic acid |
| BZ | benzodiazepine |
| CA1 | hippocampal CA1-region |
| CHA | [$^3$H]-n$^6$-Cyclohexyladenosine |
| DG | dentate gyrus |
| Ent | entorhinal cortex |
| GABA | γ-amino-butyric acid |
| GN | geniculate nuclei |
| Gpp(NH)p | guanosine 5'-[β,γ-imido]triphosphate |
| HSP-27 | heat shock protein 27 |
| NMDA | N-methyl-D-aspartate |
| ParI | somatosensory cortex |
| Pir | piriform cortex |
| PRh | perirhinal cortex |
| PTZ | Pentylenetetrazole (6,7,8,9-tetrahydro-5H-Tetrazolo(1,5-a)azepine) |
| PVP | posterior paraventricular thalamic nucleus |
| ROI | region of interest |
| RSG | retrosplenial granular cortex |
| Th | thalamus |

## Abstract

Pentylenetetrazole (PTZ) is a convulsant used to model epileptic seizures in rats. In the PTZ-model, altered heat shock protein 27 (HSP-27) expression highlights seizure affected astrocytes, which play an important role in glutamate and GABA metabolism. This raises the question whether impaired neurotransmitter metabolism leads to an imbalance in neurotransmitter receptor expression. Consequently, we investigated the effects of seizures on the densities of seven different neurotransmitter receptors in rats which were repeatedly treated with PTZ (40 mg/kg) over a period of 14 days.

Quantitative in vitro receptor autoradiography was used to measure the regional binding site densities of the glutamate α-amino-3-hydroxy-5-methyl-4-isoxazolepropionic acid (AMPA), kainate and NMDA receptors, the adenosine receptor type 1 ($A_1$), which is part of the system controlling glutamate release, and the γ-Aminobutyric acid (GABA) receptors $GABA_A$ and $GABA_B$ as well as the $GABA_A$ associated benzodiazepine (BZ) binding sites in each rat.

Our results demonstrate altered receptor densities in brain regions of PTZ-treated animals, including the HSP-27 expressing foci (i.e. amygdala, piriform and enthorinal cortex, dentate gyrus). A general decrease of kainate receptor densities was observed together with an increase of NMDA binding sites in the hippocampus, the somatosensory, piriform and the entorhinal cortices. Furthermore, $A_1$ binding sites were decreased in the amygdala and CA1, while BZ binding sites were increased in the dentate gyrus and CA1.

Our data demonstrate the impact of PTZ induced seizures on the densities of kainate, NMDA, $A_1$ and BZ binding sites in epileptic brain. These changes are not restricted to regions showing glial impairment. Thus, an altered balance between different excitatory (NMDA) and modulatory receptors ($A_1$, BZ binding sites, kainate) shows a much wider regional distribution than that of glial HSP-27 expression, indicating that receptor changes are not following the glial stress responses, but may precede the HSP-27 expression.

Keywords: epilepsy, animal model, neurotransmitter receptors, PTZ

## Introduction

Epileptic seizures are associated with an imbalance between excitatory and inhibitory neurotransmission, and may be the result of changes in the expression of not only one but various excitatory, inhibitory and/or modulatory neurotransmitter receptors.

Several studies demonstrated alterations of neurotransmitter receptors in human epileptic tissue. For example, we demonstrated an increase of AMPA binding sites in neocortical tissue of patients suffering from focal epilepsy (Zilles et al., 1999). Decreased kainate densities along with increased AMPA and NMDA binding sites have been demonstrated in temporal lobe epilepsy (Brines et al., 1997). GABA receptors were seen to be increased (Hammers et al., 2001; Bidmon et al., 2002) or decreased (Koepp et al., 1997 a,b; Bidmon et al., 2002) both in surgery explants as well as in vivo. A decrease of GABAergic receptors together with a concomitant increase of NMDA receptors was also demonstrated (Crino et al., 2001). The available data are heterogenous and it must be noted that these studies are based on different methodological approaches and stages of the disease. In addition, neurotransmitter receptors were shown to be differentially regulated during acute phase and chronic epilepsy (e.g. AMPA and NMDA receptors (Doi et al., 2001), $GABA_A$ receptors (Nobrega et al., 1990), BZ binding sites (Rocha and Ondarza-Rovira, 1999). Several animal models have been established to study acute and chronic epileptic activity as well as epileptogenesis (reviewed by Löscher, 2002; Pitkänen et al., 2007).

Intraperitoneal applications of PTZ are used to investigate the effects of acute and chronic epileptic seizures (Caspers and Speckmann, 1972; Bertram, 2007). PTZ acts as a $GABA_A$ receptor antagonist (Macdonald and Barker, 1978). Kindled animals have been shown to exhibit alterations of glutamatergic, GABAergic and adenosinergic receptors (Angelatou, 1990; Pagonopoulou et al., 1993; Ekonomou and Angelatou, 1999; Walsh et al., 1999; Ekonomou et al., 2001; Tchekalarova, 2005). An upregulation of AMPA binding was shown (Ekonomou et al., 2001), which depends on the duration of the PTZ-treatment, as well as a long lasting upregulation of NMDA binding sites (Ekonomou and Angelatou, 1999). The $GABA_A$ receptor subunit composition was transiently affected with minimal or no effect on $GABA_A$ receptor or BZ binding site densities (Walsh et al., 1999). A long-term increase of $A_1$ receptor binding in numerous regions of the brain was also reported (Angelatou et al., 1990; Pagonopoulou et al., 1993; Tchekalarova et al., 2005), but a persistent down-regulation of $A_1$ receptors was found in the basal ganglia (Pagonopoulou et al., 1993).

We previously demonstrated that heat shock protein 27 (HSP-27) is increased in human epileptic brain tissue as well as in a rat model of PTZ-induced seizures (Bidmon et al., 2004, 2005). In the PTZ model, HSP-27 labels affected astrocytes and endothelial cells (Bidmon et al., 2005, 2008). Astrocytes are key elements for the recycling and regulation of neurotransmitters, i.e. glutamate reuptake, glutamate glutamine shuttle and GABA metabolism (Martinez-Hernandez et al., 1977; Schousboe and Waagepetersen, 2006). Furthermore, astrocytes are a major source and regulator of synaptic adenosine and evidence is summing up for an important role of this neuromodulator during epiletogenesis (Gouder et al., 2004; Fedele et al., 2005; Li et al., 2007). Additionally, we recently demonstrated that astrocytic glutamine synthetase becomes nitrated and partially inhibited after repeated PTZ induced seizures (Bidmon et al., 2008). Therefore, we hypothesize that impaired astrocytes may contribute to an altered recycling of neurotransmitters, thus affecting neurotransmission after PTZ-treatment.

To test the hypothesis of an imbalance between excitatory and inhibitory neurotransmission in our PTZ-model of repeated seizures, we investigated the binding site densities of AMPA, kainate, NMDA, $A_1$, $GABA_A$, and $GABA_B$ receptors as well as of $GABA_A$ associated benzodiazepine binding sites in different brain regions of each PTZ-treated rat. A further goal was to elucidate whether the localization of changes in receptor densities corresponds to that seen for the HSP-27 glial stress response (Bidmon et al., 2005). Therefore, regions of interest were chosen according to prior observations of a strong HSP-27 response, and further brain regions relevant for the pathology of induced seizures were studied where no significant HSP-27 increase had been found previously.

## Experimental Procedures

Animals
We used 8-week-old male Wistar rats (220-250 g bodyweight, Moellegaard Breeding Centre GmbH, Germany) for all experiments, and animals were housed under standard conditions as previously described (Rauca et al., 2004, Bidmon et al., 2005). All experimental procedures were conducted according to the institutional guidelines for the use of laboratory animals, the German Animal Welfare Act and approved by the responsible governmental agency.

Pentylenetetrazole treatment
Rats were injected intraperitoneally with PTZ (40 mg/kg bodyweight) dissolved in physiological saline (20 mg/ml), leading to acute tonic-clonic seizures. Epileptic activity was logged as follows: Four rats per group were kept in cages with an assembled recording setup to monitor seizure activity and duration (Bidmon et al., 2005), while the seizure intensity of all other animals was estimated in a scoring system according to Rauca et al. (2004). Continuous treatment (every 48h for 14 days) induced periodic epileptic activity as described previously (Caspers and Speckmann, 1972; Rauca et al., 2004). All rats repeatedly exhibtited tonic-clonic seizures at the latest from the fourth treatment and were used for receptor autoradiography (n = 8). Control rats (n = 6) were treated with physiological saline with the same frequency as PTZ-treated rats. Animals were decapitated 24h after the last treatment and brains immediately frozen in isopentane at -50 °C.

[$^3$H]-Receptor autoradiography
Unfixed, deep-frozen brains were serially sectioned (20 µm) in the coronal plane with a Cryostat (Leica, Germany). Alternating sections mounted on glass slides were used for quantitative *in vitro* receptor autoradiography (Zilles et al., 2002; Zilles et al., 2004; Palomero-Gallagher et al., 2008) or cell body staining (Merker, 1983). In brief, sections were washed in buffer and subsequently incubated in buffer containing the respective [$^3$H]-ligand (Table 1). Nonspecific binding was monitored by incubation in the presence of a specific, non-radioactive competitor (Table 1). Finally, sections were rinsed in buffer. Experimental and control brains were processed simultaneously. Sections were exposed to ß-sensitive films (BioMax MR Film, Kodak Europe) for 10 to 15 weeks in the presence of standards of known radioactivity concentrations (Amersham Biosciences Europe). All ligands were purchased from

Perkin Elmer (Germany) except for [$^3$H]-CGP 54626 (BioTrend, Germany). Competitors R-PIA and Clonazepam were purchased from Sigma Aldrich (Germany), CGP 55845 and SYM 2081 from Tocris Bioscience (Great Britain) and quisqualate, MK-801 and GABA from Biotrend (Germany).

Image analysis

Binding site densities (fmol/mg protein) were measured densitometrically as described in detail (Zilles and Schleicher, 1991, 1995; Zilles et al., 2002). After development of the autoradiographic films (Hyperprocessor, Amersham Biosciences Europe), they were digitized (8-bit coding, 1300 x 1030 pixels) using a CCD-camera (Progres C14, Jenoptik, Germany) equipped with a 55 mm lens (Zeiss, Germany) and the KS400 image analysing software (Zeiss, Germany). The standards were used to compute a transformation curve indicating the relationship between grey values in the autoradiographs and concentrations of radioactivity in the tissue. The digitized autoradiographs were printed, and regions of interest (ROIs) were traced according to cytoarchitectonic criteria (Zilles, 1985) by comparison with neighbouring histological sections and a drawing microscope (Axioscope, Zeiss, Germany). The grey values in the cytoarchitectonically defined ROIs were retrieved using a graphics tablet (Wacom, Germany). Regional receptor densities were calculated in various cortical regions. In each animal and ROI, receptor densities were measured in 3-5 randomised sections.

For Figures 1-2, autoradiographs were additionally colour-coded with the KS400 image analysing software solely to provide a clear visual impression of the regional distribution of receptor binding sites. The 256 grey values of each linearized autoradiograph were colour-coded by assigning eleven colours in a spectral sequence to equally spaced grey value ranges.

Statistical analysis

An analysis of variance with repeated measurements was performed to determine significant differences of regional mean receptor binding site densities between experimental and control animals ($p \leq 0.05$ or $p \leq 0.01$).

Table 1: Brief summary of [³H]-ligands, displacers and binding conditions of autoradiographic experiments. *only added in main incubation; **only added in pre- and main incubation

| Receptor | [³H]-Ligand | Displacer | Incubation buffer | Pre-incubation | Main incubation | Rinsing |
|---|---|---|---|---|---|---|
| AMPA | AMPA, 10 nM | Quisqualate, 10 μM | 50 mM Tris-acetate (pH: 7.2) + 100 mM KSCN* | 3x10 min at 4°C | 45 min at 4°C | 4x4 sec in buffer at 4°C |
| Kainate | Kainate, 9.4 nM | SYM 2081, 100 μM | 50 mM Tris-citrate (pH: 7.1) + 10 mM Calcium acetate* | 3x10 min at 4°C | 45 min at 4°C | 2x2 sec in 2.5% glutaraldehyde in acetone 3x4 sec at 4°C |
| NMDA | MK 801, 3.3 nM | MK 801, 100 μM | 50 mM Tris-HCl (pH: 7.2) + 50 μM Glutamate** + 30 μM Glycine* + 50 μM Spermidine* | 15 min at 4°C | 60 min, 20°C | 2x2 sec in 2.5% glutaraldehyde in acetone 2x5 min at 4°C 1 sec in a. dest. |
| GABA_A | Muscimol, 7.7 nM | GABA, 10 μM | 50 mM Tris-citrate (pH: 7.0) | 3x5 min at 4°C | 40 min at 4°C | 3x3 sec at 4°C 1 sec. in a. dest. |
| GABA_B | CGP 54626, 2 nM | CGP 55845, 100 μM | 50 mM Tris-HCl (pH: 7.2) + 2.5 mM Calcium chloride | 3x5 min at 4°C | 60 min at 4°C | 3x2 sec at 4°C 1 sec in a. dest. at 4°C |
| BZ binding site | Flumazenil, 1 nM | Clonazepam, 2 μM | 170 mM Tris-HCl (pH: 7.4) | 15 min at 4°C | 60 min at 4°C | 1 sec in a. dest. at 4°C 2x1 min at 4°C |
| A_1 | DPCPX + Gpp(NHp), 1 nM | R-PIA, 100 μM | 170 mM Tris-HCl (pH: 7.4) + 2 Units/l Adenosine deaminase + 100μM Gpp(NH)p* | 15 min at 4°C | 120 min, 20°C | 2x5 min at 4°C 1 sec in a. dest. at 4°C |

## Results

In PTZ-treated and control rats, mean densities (fmol/mg protein) of AMPA, kainate, NMDA, $A_1$, $GABA_A$, BZ and $GABA_B$ binding sites were measured in the amygdala, the piriform, entorhinal, somatosensory and retrosplenial granular cortices as well as in the hippocampal CA1 region and the dentate gyrus (Fig. 1, 2; Tab. 2). Significant differences between control and PTZ-treated rats were found for kainate, NMDA, BZ and $A_1$ binding sites (Fig. 3).

High densities of AMPA receptors were observed in neocortical as well as hippocampal regions of both control and experimental animals (Fig. 1 C, D). In the neocortex, AMPA receptor densities were more prominent in supragranular (superficially of layer IV) than in infragranular layers. Highest regional densities were measured in the hippocampal CA1 subfield and the dentate gyrus of control as well as PTZ-treated animals. No significant differences in the binding site densities of AMPA receptors were found between PTZ-treated and control animals.

In contrast to AMPA receptors, kainate binding sites exhibited high densities in the infragranular cortical layers V and VI as well as in the putamen, CA3 region and the dentate gyrus. By far the highest kainate receptor densities were found in the stratum lucidum of CA3 and in the hilus region. A general decrease in kainate binding sites of 20-40 % was found in all regions analyzed of PTZ-treated rats compared to controls (Fig. 3).

The laminar distribution of NMDA receptors was similar to that of AMPA receptors (Fig. 1 G, H). Compared to AMPA receptors, a very low density of NMDA receptors was found in the pyramidal layer of CA1-3, but high densities in subcortical areas like the thalamus (Fig. 1 G) and the geniculate nuclei (Fig. 1 H). A general increase of NMDA binding in PTZ-treated animals was already seen in the colour coded autoradiographs. A comparison between PTZ-animals and the controls showed a significant increase of binding site densities (20-25 %) in the somatosensory, piriform and entorhinal cortices as well as in the dentate gyrus and the CA1 region in the PTZ-treated group, while the increase in the amygdala and retrosplenial granular cortex did not reach significance (Fig. 3).

Highest densities of $A_1$ receptors were seen in the hippocampal CA1 region and the dentate gyrus, as well as in subcortical areas, e.g. thalamic and geniculate nuclei (Fig. 2 G, H). The medial and dorsal neocortex displayed higher densities than ventral regions, i.e. perirhinal, piriform or entorhinal cortex (Fig. 2 A, B). Highest $A_1$

densities were found in the infragranular layers. In PTZ-rats, a significant decrease (17-20 %) of $A_1$ receptor binding site densities was found in the amygdala and the hippocampal CA1 region when compared to controls (Fig. 3). Contrast enhanced and colour-coded images also suggest decreased $A_1$ receptor densities in the thalamus of PTZ-treated rats (Fig. 2 A), but measurements did not reveal significant differences between control (945 ± 43 fmol/mg protein ± S.E.M) and PTZ-treated rats (869 ± 33 fmol/mg protein ± S.E.M).

In both the control and experimental group, high densities of $GABA_A$ receptors were found in the neocortex as well as in subcortical nuclei (Fig. 2 C, D). The hippocampus, retrosplenial cortex and the amygdala exhibited low densities. In the hippocampus, $GABA_A$ receptors were most prominent in CA1 and in the dentate gyrus, with decreasing binding site densities along the rostro-caudal axis. CA2, CA3 and the hilar region showed very low densities. No significant differences in $GABA_A$ receptor binding site densities could be found between control and PTZ-treated rats in any of the analyzed brain regions.

BZ binding site densities were high in cortical areas and low in subcortical areas (Fig. 2 E, F). Within the neocortex, highest BZ binding site densities were present in layer V. Very high densities were also observed in the dentate gyrus. The lowest densities were measured in the pyramidal layer of CA2-3. Significantly increased (16-17 %) BZ binding sites in PTZ-treated rats compared to controls were found in the dentate gyrus and CA1 region (Fig. 3).

$GABA_B$ receptors occur at high densities in the thalamic nuclei, particularly in the paraventricular thalamic nucleus and the medial and lateral geniculate nuclei (Fig. 2, G, H). In neocortical areas the highest densities of $GABA_B$ receptors were found in the supragranular layers. In the insular, piriform or entorhinal regions, these receptors appeared to be more homogenously distributed. Within the hippocampal formation, the highest densities of $GABA_B$ receptors were measured in the dentate gyrus, but only low amounts throughout CA1. The pyramidal layer as well as the hilar region exhibited very low receptor densities. We did not find any significant differences in density of $GABA_B$ receptors between control and PTZ-treated animals.

Figure 1: Colour coded autoradiographs of AMPA, kainate and NMDA receptors in brains of control and PTZ-treated rats. Scale bars code receptor densities in fmol/mg protein. Within each receptor type, demonstrated hemispheres were colour coded using the same scaling. (A) Cytoarchitectonic definition of measured regions (grey) at Bregma -3.3 mm and (B) at Bregma -5.3 mm according to Zilles (1985). For detailed description see text. (C-D) AMPA receptors, (E-F) kainate receptors, (G-H) NMDA receptors. *Amygdala (AMG); piriform cortex (Pir); entorhinal cortex (Ent); somatosensory cortex (Parl); retrosplenial granular cortex (RSG); dentate gyrus (DG);hippocampal CA1-region (CA1); posterior paraventricular thalamic nucleus (PVP); thalamus (Th); geniculate nuclei (GN); perirhinal cortex (PRh)*

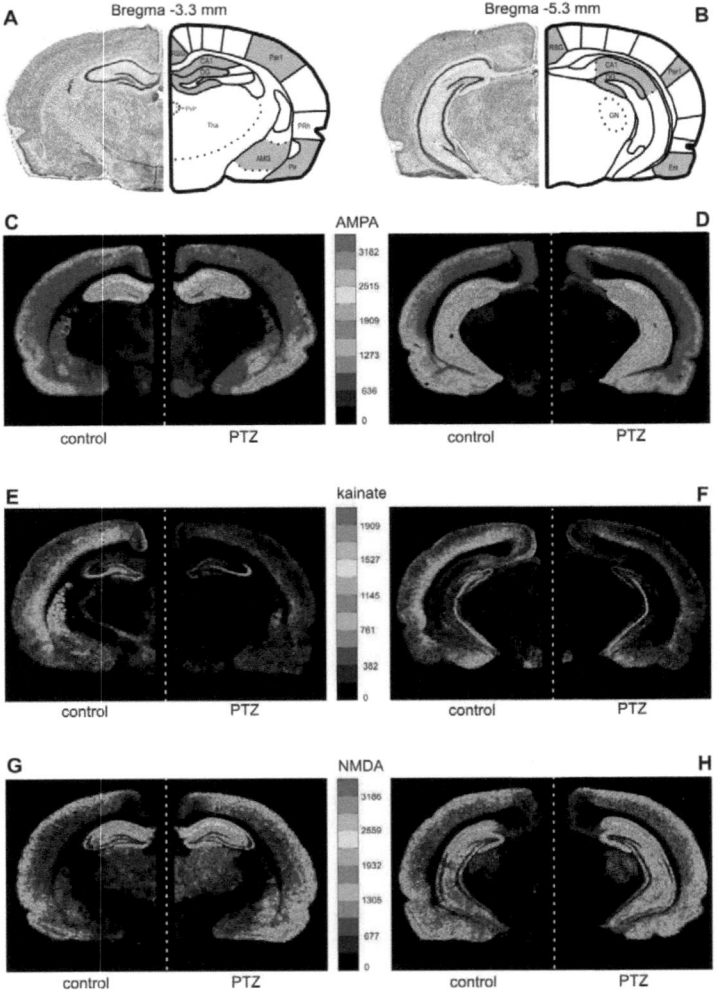

Figure 2: Colour coded autoradiographs of $A_1$, $GABA_A$, BZ and $GABA_B$ binding sites in brains of control and PTZ-treated rats. Scale bars code receptor densities in fmol/mg protein. Within each receptor type, demonstrated hemispheres were colour coded using the same scaling. A cytoarchitechtonic definition of measured regions is demonstrated in figure 1 A-B. For detailed description see text. (A-B) $A_1$ binding sites, (C-D) $GABA_A$ receptors, (E-F) BZ binding sites, (G-H) $GABA_B$ receptors.

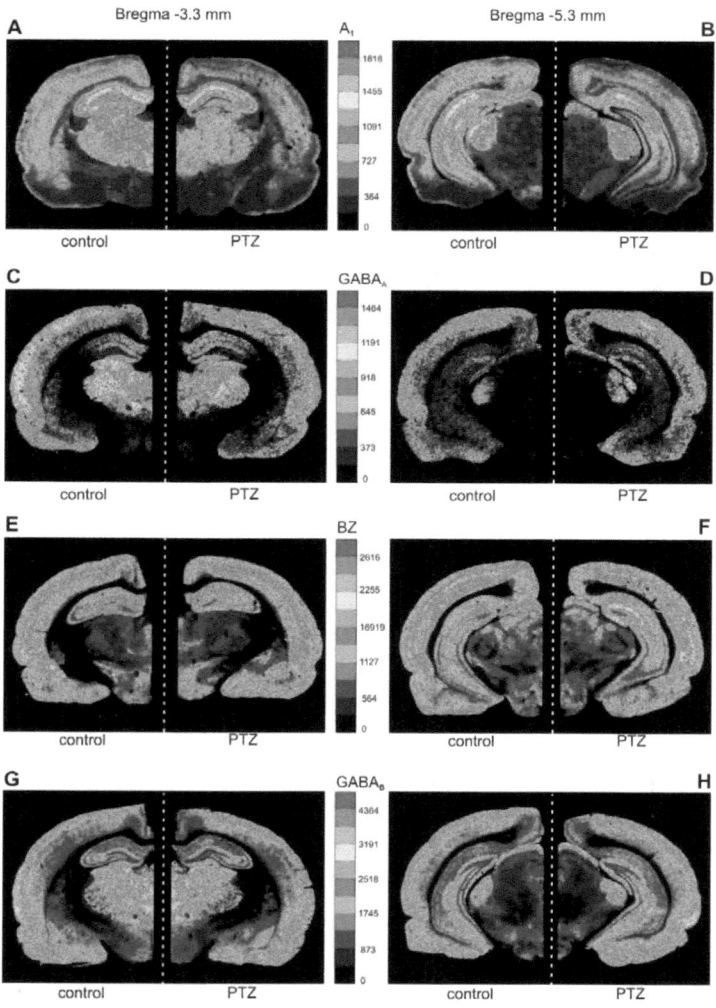

Table 2: Mean receptor binding site densities in fmol/mg protein ± S.E.M. measured in the brains of control (n = 6) and PTZ-treated (n = 8) rats. Significant differences are marked by one (p < 0.05) or two (p < 0.01) asterisks. Amygdala (**AMG**); piriform cortex (PIR); entorhinal cortex (ENT); somatosensory cortex (ParI); retrosplenial granular cortex (RSG); dentate gyrus (DG); hippocampal CA1-region (CA1)

| Receptor | Treatment | Amygdala | Piriform | Entorhinal | ParI | RSG | Mol | CA1 |
|---|---|---|---|---|---|---|---|---|
| AMPA | PTZ | 739 ± 22 | 1324 ± 45 | 1232 ± 37 | 780 ± 27 | 674 ± 26 | 1599 ± 94 | 1897 ± 144 |
|  | Control | 770 ± 40 | 1366 ± 42 | 1343 ± 102 | 877 ± 10 | 806 ± 40 | 1946 ± 107 | 2028 ± 49 |
| Kainate | PTZ | 329** ± 17 | 341** ± 13 | 326** ± 10 | 274** ± 20 | 222** ± 20 | 338** ± 12 | 125** ± 13 |
|  | Control | 459 ± 15 | 411 ± 7 | 422 ± 18 | 392 ± 11 | 352 ± 18 | 418 ± 16 | 213 ± 11 |
| NMDA | PTZ | 611 ± 41 | 1281** ± 43 | 1354** ± 59 | 930** ± 30 | 550 ± 23 | 1678** ± 73 | 1897** ± 114 |
|  | Control | 565 ± 27 | 1032 ± 50 | 1105 ± 39 | 776 ± 43 | 528 ± 54 | 1363 ± 50 | 1512 ± 130 |
| GABA$_A$ | PTZ | 590 ± 47 | 1291 ± 69 | 1268 ± 48 | 1397 ± 73 | 1039 ± 83 | 1554 ± 63 | 1045 ± 111 |
|  | Control | 480 ± 28 | 1095 ± 76 | 1244 ± 106 | 1359 ± 38 | 1012 ± 76 | 1178 ± 24 | 937 ± 51 |
| GABA$_B$ | PTZ | 1913 ± 20 | 2158 ± 35 | 2276 ± 51 | 1549 ± 21 | 1155 ± 17 | 1701 ± 36 | 1202 ± 25 |
|  | Control | 1796 ± 122 | 1963 ± 93 | 2184 ± 55 | 1564 ± 53 | 1235 ± 43 | 1791 ± 42 | 1260 ± 40 |
| A$_1$ | PTZ | 308** ± 11 | 399 ± 14 | 371 ± 16 | 749 ± 46 | 682 ± 28 | 839 ± 27 | 1136* ± 42 |
|  | Control | 382 ± 12 | 437 ± 14 | 395 ± 17 | 788 ± 28 | 770 ± 40 | 859 ± 38 | 1364 ± 77 |
| BZ | PTZ | 1005 ± 22 | 1225 ± 25 | 1229 ± 33 | 1371 ± 46 | 1176 ± 48 | 1674 ± 39 | 1400 ± 49 |
|  | Control | 935 ± 44 | 1232 ± 29 | 1164 ± 42 | 1354 ± 50 | 1146 ± 66 | 1437 ± 54 | 1193 ± 73 |

Figure 3: Bar charts of regional receptor binding site densities in fmol/mg protein ± S.E.M of kainate (A), NMDA (B), BZ (C) and $A_1$ (D) in control (n = 6) and PTZ-treated (n = 8) rats. Significant differences are marked by one ($p \leq 0.05$) or two ($p \leq 0.01$) asterisks. *Amygdala (AMG); piriform cortex (PIR); entorhinal cortex (ENT); somatosensory cortex (Parl); retrosplenial granular cortex (RSG); dentate gyrus (DG); hippocampal CA1-region (CA1).*

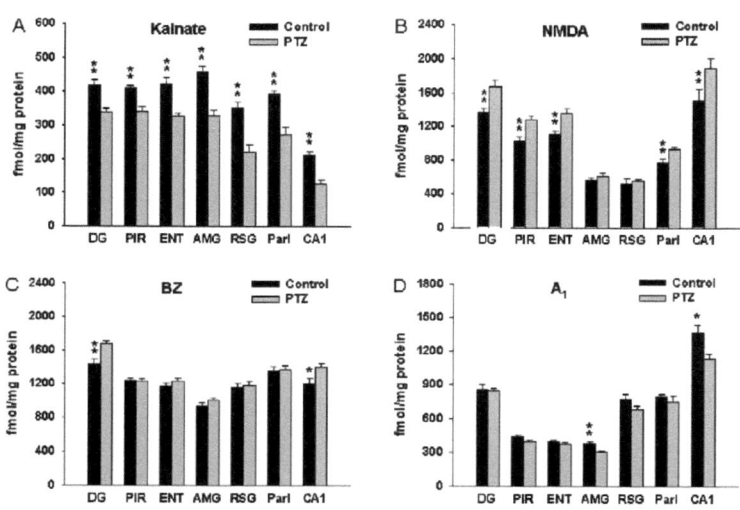

## Discussion

The aim of this study was to determine the effects of repeated PTZ induced seizures on the densities of binding sites of different excitatory and inhibitory neurotransmitter receptors in the adult rat brain. A reduction of kainate receptor binding sites, but an increase of NMDA receptors as well as of BZ binding sites were found in most brain regions studied. The density of $A_1$ receptor binding sites was reduced in a few regions.

PTZ was introduced as a $GABA_A$ antagonist (Macdonald and Barker, 1978), and later demonstrated to bind at BZ and picrotoxin binding sites (Rehavi et al., 1982; Squires et al., 1984; Huang et al. 2001). Thus, it can be hypothesized that the PTZ-treatment may reduce the binding to BZ binding sites of $GABA_A$ receptors, but we did not find such reductions in the present study. In contrast, we found an increase of BZ binding site densities in the CA1 region and the dentate gyrus of PTZ-treated rats. This result can be explained by alterations of the subunit-composition of the functional $GABA_A$ receptor complex in these areas, since binding sites for GABA and BZ have different localisation and structural properties at the functional receptor (Pritchett et al., 1989; Rudolph et al., 1999; McKernan et al., 2000; Baumann et al., 2003). Binding of BZ is known to enhance $GABA_A$ receptor activity. It was demonstrated that $GABA_A$ agonist administration reduces the effect after a single PTZ application as well as after chronic PTZ treatment (Hansen et al., 2004). Therefore, the increase of BZ binding sites might attenuate neuronal hyperactivity in regions which have been shown to be particularly susceptible to epileptic activity (Majores et al., 2007; Scharfman and Gray, 2007).

Autoradiographic studies in mice demonstrated a gradual and long-lasting increase of NMDA receptor binding during a 26 day process of PTZ-kindling in the dentate gyrus and CA3 together with a short term increase in the somatosensory cortex (Ekonomou and Angelatou, 1999). Here, we demonstrated that NMDA receptor up-regulation in CA1 and dentate gyrus, as well as in the somatosensory, piriform and entorhinal cortices occurs already after two weeks. Thus, these findings support the hypothesis that an increased NMDA receptor density is caused by PTZ-induced seizure activity. Rats exposed to moderate hypoxia exhibit a reduced susceptibility to PTZ-induced seizures (Rauca and Ruethrich, 1995). Since application of NMDA or $A_1$ receptor antagonists before hypoxia blocked the decrease of PTZ-susceptibility, an involvement of these receptors has been proposed in the underlying neuroprotective mechanism. However, the inhibition of NMDA receptors has also been shown to reduce PTZ-induced seizure activity (Jiang et al., 2004).

$A_1$ receptor activation can mediate antiepileptic effects in the PTZ-model (Barraco, 1984; Malhotra and Gupta, 1997; Zgodzinsdki et al., 2001; Omrani and Fathollahi, 2003; Ates et al., 2005). Activation of $A_1$ receptors results in a reduction of glutamate release (Prestwich et al., 1987) and modulates neuronal function (Fredholm et al., 2005). There are several reports of $A_1$ receptor densities being upregulated in brain regions of PTZ-treated mice, including the amygdala and the CA1 region (Angelatou et al., 1990; Pagonopoulou et al., 1993; Tchekalarova et al., 2005). In contrast to these reports, we observed significantly decreased $A_1$ receptor binding site densities in these regions.

Synaptic adenosine release is differentilally regulated during acute and chronic seizure situations. This is mainly due to a biphasic expression of the astrocytic enzyme adenosine kinase (for review see Boison, 2008). Since we demonstrated enhanced affection of astrocytes in our PTZ-model of repeatedly induced seizures, it may be hypothesized that our results represent a chronic epileptic situation. However, we did not observe evidence of enhanced cellular decline, which is a hallmark of chronic epileptic tissue. According to electrophysiologal experiments, our treatment serves as a model of repeated seizure episodes, since to achieve fully kindled animals PTZ-treatment must continue for longer periods or must be applied in shorter intervals (e.g. Rauca et al., 2004; Park et al., 2006).

It seems likely that the results described above are in fact not contradictory, but the result of differences in the methodological approach. We used the inverse agonist [$^3$H]-DPCPX in the presence of guanosine 5'-[β,γ-imido]triphosphate (Gpp(NH)p), a stable GTP-analog that decouples the G-protein from the receptor. Therefore, the presence of Gpp(NH)p during incubation enables binding to all available $A_1$ receptors independent from their state of activation. The studies noted above were performed using the agonistic [$^3$H]-n$^6$-Cyclohexyladenosine (CHA) in absence of Gpp(NH)p. Since binding of CHA to $A_1$ receptors is highly dependent on the status of G-protein coupling (Fastbom and Fredholm, 1990), it seems likely that the discussed results are in fact not contradictory, but the effect of differences in the methodological approach. Future effort is necessary to elucidate this subject.

Kainate receptors are generally believed to play a role in the epileptic pathology. However, due to their ubiquitous occurrence, varying synaptic and subcellular localisations as well as their multiple functions (Lerma, 2003; Pinheiro and Mulle, 2006), the involvement of kainate receptors in epileptic seizures remains a matter of debate. We demonstrated decreased kainate binding sites in all investigated regions of PTZ-treated rats. In accordance with our findings, it has been described that fully PTZ-kindled rats (30 mg/kg, 4-6 weeks daily) exhibit reduced kainate binding (Luthman and Humpel, 1997). It remains to be addressed though, whether PTZ-treatment directly affects receptor expression or whether the observed decrease is secondary to epileptic activity. A follow up study using lower clearly subconvulsive doses of PTZ would complement our present data.

HSP-27 expression is regionally elevated in astrocytes due to oxidative stress in both human epileptic patients as well as in the PTZ-model (Bidmon et al., 2004, 2008). Therefore, one goal of the present study was to test the hypothesis that PTZ-induced impairment of glia cells results in alterations of neurotransmitter receptor densities. We found a partial overlap of regions showing glial heat shock response and altered receptor densities. Regions in which altered HSP-27 expression and binding site densities coincided were the piriform-entorhinal cortex, the dentate gyrus and the amygdala. In the somatosensory cortex receptor binding sites were significantly changed, but not the glial heat shock response (Bidmon et al., 2005). For the retrosplenial cortex we found a decrease of kainate binding but no increase in HSP-27 induction (Bidmon et al. 2005). Therefore, changes in the retrosplenial cortex would correspond to the early activations observed during PTZ-induced seizures as revealed by functional imaging (Brevard et al. 2006) or c-fos induction (André et al., 1998). However, the most striking difference was detected in CA1, where considerable alterations of receptor binding sites were not accompanied by a glial heat shock response. We therefore suggest, that alterations of receptor densities in the PTZ-model occur independent of corresponding glial impairment and precede the glial reaction in a region-specific manner. It has to be noted though, that region-specific differences among astrocytes have been reported. For example CA1 glia cells are characterized by a clear lack of NMDA receptors and ionotropic receptors for extracellular adenosine (Matthias et al., 2003; Lalo et al., 2006; Jabs et al., 2007). These differences may contribute to the fact that some regions show receptor alterations and concomitant glial changes, whereas in other regions receptor differences occur without severe and already detectable concomitant glial responses.

Taken together, our results demonstrate alterations of kainate, NMDA, $A_1$ and

BZ binding site densities in epileptic tissue of our PTZ-model. These changes were independent of glial impairment as indicated by comparison with HSP-27 expression. Further effort is necessary to elucidate whether PTZ directly influences receptor expression, or whether these changes are secondary to epileptic onset.

**Acknowledgments:**

The authors appreciate the technical support by M. Cremer, S. Krause and S. Wilms. We would also like to thank A. Bauer for helpful comments on adenosine receptor properties.

## Reference List

Angelatou F, Pagonopoulou O, Kostopoulos G (1990) Alterations of $A_1$ adenosine receptors in different mouse brain areas after pentylentetrazol-induced seizures, but not in the epileptic mutant mouse 'tottering'. Brain Res 534:251-256.

André V, Pineau N, Motte JE, Marescaux C, Nehlig A (1998) Mapping of neuronal networks underlying generalized seizures induced by increasing doses of pentylenetetrazol in the immature and adult rat: a c-Fos immunohistochemical study. Eur J Neurosci. 10(6):2094-106.

Ates N, Ilbay G, Sahin D (2005) Suppression of generalized seizures activity by intrathalamic 2-chloroadenosine application. Exp Biol Med (Maywood) 230:501-505.

Barraco RA, Swanson TH, Phillis JW, Berman RF (1984) Anticonvulsant effects of adenosine analogues on amygdaloid-kindled seizures in rats. Neurosci Lett 46:317-322.

Baumann SW, Baur R, Sigel E (2003) Individual properties of the two functional agonist sites in $GABA_A$ receptors. J Neurosci 23:11158-11166.

Bertram E (2007) The relevance of kindling for human epilepsy. Epilepsia 48 Suppl 2:65-74.

Bidmon HJ, Görg B, Palomero-Gallagher N, Behne F, Lahl R, Pannek HW, Speckmann EJ, Zilles K (2004) Heat shock protein-27 is upregulated in the temporal cortex of patients with epilepsy. Epilepsia 45:1549-1559.

Bidmon HJ, Görg B, Palomero-Gallagher N, Schleicher A, Häussinger D, Speckmann EJ, Zilles K (2008) Glutamine synthetase becomes nitrated and its activity is reduced during repetitive seizure activity in the pentylentetrazole model of epilepsy. Epilepsia 49(10):1733-1748

Bidmon HJ, Gorg B, Palomero-Gallagher N, Schliess F, Gorji A, Speckmann EJ, Zilles K (2005) Bilateral, vascular and perivascular glial upregulation of heat shock protein-27 after repeated epileptic seizures. J Chem Neuroanat 30:1-16.

Bidmon HJ, Palomero-Gallagher N, Zilles K (2002) Postoperative Untersuchungen in der Epilepsiechirurgie: Enzyme der oxidativen Stresskaskade, Multi-Drug-Transporter und Transmitterrezeptoren. Klin Neurophysiol 33:168-177.

Boison D (2008) Adenosine as a modulator of brain activity. Drug News Perspect. 20(10):607-11.

Brevard ME, Kulkarni P, King JA, Ferris CF (2006) Imaging the neural substrates involved in the genesis of pentylenetetrazol-induced seizures. Epilepsia. 47(4):745-54.

Brines ML, Sundaresan S, Spencer DD, de Lanerolle NC (1997) Quantitative autoradiographic analysis of ionotropic glutamate receptor subtypes in human temporal lobe epilepsy: up-regulation in reorganized epileptogenic hippocampus. Eur J Neurosci. 9(10):2035-44.

Caspers H, Speckmann EJ (1972) Cerebral $pO_2$, $pCO_2$ and pH: changes during convulsive activity and their significance for spontaneous arrest of seizures. Epilepsia 13:699-725.

Crino PB, Duhaime AC, Baltuch G, White R (2001) Differential expression of glutamate and $GABA_A$ receptor subunit mRNA in cortical dysplasia. Neurology 56:906-913.

Doi T, Ueda Y, Tokumaru J, Mitsuyama Y, Willmore LJ (2001) Sequential changes in AMPA and NMDA protein levels during $Fe^{3+}$-induced epileptogenesis. Brain Res Mol Brain Res 92:107-114.

Ekonomou A, Angelatou F (1999) Upregulation of NMDA receptors in hippocampus and cortex in the pentylenetetrazol-induced "kindling" model of epilepsy. Neurochem Res 24:1515-1522.

Ekonomou A, Smith AL, Angelatou F (2001) Changes in AMPA receptor binding and subunit messenger RNA expression in hippocampus and cortex in the pentylenetetrazole-induced 'kindling' model of epilepsy. Brain Res Mol Brain Res 95:27-35.

Fastbom J, Fredholm BB (1990) Regional differences in the effect of guanine nucleotides on agonist and antagonist binding to adenosine $A_1$-receptors in rat brain, as revealed by autoradiography. Neuroscience 34:759-769.

Fedele DE, Gouder N, Güttinger M, Gabernet L, Scheurer L, Rülicke T, Crestani F, Boison D (2005) Astrogliosis in epilepsy leads to overexpression of adenosine kinase, resulting in seizure aggravation. Brain. 2005 128(Pt 10):2383-95.

Fredholm BB, Chen JF, Masino SA, Vaugeois JM (2005) Actions of adenosine at its receptors in the CNS: insights from knockouts and drugs. Annu Rev Pharmacol Toxicol 45:385-412.

Gouder N, Scheurer L, Fritschy JM, Boison D (2004) Overexpression of adenosine kinase in epileptic hippocampus contributes to epileptogenesis. J Neurosci. 24(3):692-701.

Hammers A, Koepp MJ, Labbe C, Brooks DJ, Thom M, Cunningham VJ, Duncan JS (2001) Neocortical abnormalities of [$^{11}$C]-flumazenil PET in mesial temporal lobe epilepsy. Neurology 56:897-906.

Hansen SL, Sperling BB, Sanchez C (2004) Anticonvulsant and antiepileptogenic effects of $GABA_A$ receptor ligands in pentylenetetrazole-kindled mice. Prog Neuropsychopharmacol Biol Psychiatry 28:105-113.

Huang RQ, Bell-Horner CL, Dibas MI, Covey DF, Drewe JA, Dillon GH (2001) Pentylenetetrazole-induced inhibition of recombinant gamma-aminobutyric acid type A ($GABA_A$) receptors: mechanism and site of action. J Pharmacol Exp Ther 298:986-995.

Jabs R, Matthias K, Grote A, Grauer M, Seifert G, Steinhauser C (2007) Lack of P2X receptor mediated currents in astrocytes and GluR type glial cells of the hippocampal CA1 region. Glia 55:1648-1655.

*Jiang W, Wolfe K, Xiao L, Zhang ZJ, Huang YG, Zhang X (2004) Ionotropic glutamate receptor antagonists inhibit the proliferation of granule cell precursors in the adult brain after seizures induced by pentylenetrazol. Brain Res. 1020(1-2):154-60.*

Koepp MJ, Labbe C, Richardson MP, Brooks DJ, Van PW, Cunningham VJ, Duncan JS (1997a) Regional hippocampal [$^{11}$C]flumazenil PET in temporal lobe epilepsy with unilateral and bilateral hippocampal sclerosis. Brain 120 ( Pt 10):1865-1876.

Koepp MJ, Richardson MP, Brooks DJ, Cunningham VJ, Duncan JS (1997b) Central benzodiazepine/gamma-aminobutyric acid A receptors in idiopathic generalized epilepsy: an [$^{11}$C]flumazenil positron emission tomography study. Epilepsia 38:1089-1097.

Lalo U, Pankratov Y, Kirchhoff F, North RA, Verkhratsky A (2006) NMDA receptors mediate neuron-to-glia signaling in mouse cortical astrocytes. J Neurosci 26:2673-2683.

Lerma J (2003) Roles and rules of kainate receptors in synaptic transmission. Nat Rev Neurosci 4:481-495.

Li T, Quan Lan J, Fredholm BB, Simon RP, Boison D (2007) Adenosine dysfunction in astrogliosis: cause for seizure generation? Neuron Glia Biol. 3(4):353-66.

Löscher W (2002) Animal models of epilepsy for the development of antiepileptogenic and disease-modifying drugs. A comparison of the pharmacology of kindling and post-status epilepticus models of temporal lobe epilepsy. Epilepsy Res. 50(1-2):105-23.

Luthman J, Humpel C (1997) Pentylenetetrazol kindling decreases N-methyl-D-aspartate and kainate but increases gamma-aminobutyric acid-A receptor binding in discrete rat brain areas. Neurosci Lett. 239:9-12

Macdonald RL, Barker JL (1978) Specific antagonism of GABA-mediated postsynaptic inhibition in cultured mammalian spinal cord neurons: a common mode of convulsant action. Neurology 28:325-330.

Majores M, Schoch S, Lie A, Becker AJ (2007) Molecular neuropathology of temporal lobe epilepsy: complementary approaches in animal models and human disease tissue. Epilepsia 48 Suppl 2:4-12.

Malhotra J, Gupta YK (1997) Effect of adenosine receptor modulation on pentylenetetrazole-induced seizures in rats. Br J Pharmacol 120:282-288.

Martinez-Hernandez A, Bell KP, Norenberg MD (1977) Glutamine synthetase: glial localization in brain. Science 195:1356-1358.

Matthias K, Kirchhoff F, Seifert G, Huttmann K, Matyash M, Kettenmann H, Steinhauser C (2003) Segregated expression of AMPA-type glutamate receptors and glutamate transporters defines distinct astrocyte populations in the mouse hippocampus. J Neurosci 23:1750-1758.

McKernan RM, Rosahl TW, Reynolds DS, Sur C, Wafford KA, Atack JR, Farrar S, Myers J, Cook G, Ferris P, Garrett L, Bristow L, Marshall G, Macaulay A, Brown N, Howell O, Moore KW, Carling RW, Street LJ, Castro JL, Ragan CI, Dawson GR, Whiting PJ (2000) Sedative but not anxiolytic properties of benzodiazepines are mediated by the $GABA_A$ receptor alpha1 subtype. Nat Neurosci 3:587-592.

Merker B (1983) Silver staining of cell bodies by means of physical development. J Neurosci Methods 9:235-241.

Nobrega JN, Kish SJ, Burnham WM (1990) Regional brain [3H]muscimol binding in kindled rat brain: a quantitative autoradiographic examination. Epilepsy Res. 6(2):102-9.

Omrani A, Fathollahi Y (2003) Reversal of pentylenetetrazol-induced potentiation phenomenon by theta pulse stimulation in the CA1 region of rat hippocampal slices. Synapse 50:83-94.

Pagonopoulou O, Angelatou F, Kostopoulos G (1993) Effect of pentylentetrazol-induced seizures on $A_1$ adenosine receptor regional density in the mouse brain: a quantitative autoradiographic study. Neuroscience 56:711-716.

Palomero-Gallagher N, Schleicher A, Lindemann S, Lessenich A, Zilles K, Löscher W (2008) Receptor fingerprinting the circling ci2 rat mutant: Insights into brain asymmetry and motor control. Exp Neurol. 210(2):624-37

Park JH, Cho H, Kim H, Kim K (2006) Repeated brief epileptic seizures by pentylenetetrazole cause neurodegeneration and promote neurogenesis in discrete brain regions of freely moving adult rats. Neuroscience.140(2):673-84

Pinheiro P, Mulle C (2006) Kainate receptors. Cell Tissue Res 326:457-482.

Pitkänen A, Kharatishvili I, Karhunen H, Lukasiuk K, Immonen R, Nairismägi J, Gröhn O, Nissinen J (2007) Epileptogenesis in experimental models. Epilepsia. 48 Suppl 2:13-20.

Prestwich SA, Forda SR, Dolphin AC (1987) Adenosine antagonists increase spontaneous and evoked transmitter release from neuronal cells in culture. Brain Res 405:130-139.

Pritchett DB, Luddens H, Seeburg PH (1989) Type I and type II $GABA_A$-benzodiazepine receptors produced in transfected cells. Science 245:1389-1392.

Rauca C, Ruthrich HL (1995) Moderate hypoxia reduces pentylenetetrazol-induced seizures. Naunyn Schmiedebergs Arch Pharmacol 351:261-267.

Rauca C, Wiswedel I, Zerbe R, Keilhoff G, Krug M (2004) The role of superoxide dismutase and alpha-tocopherol in the development of seizures and kindling induced by pentylenetetrazol - influence of the radical scavenger alpha-phenyl-N-tert-butyl nitrone. Brain Res 1009:203-212.

Rehavi M, Skolnick P, Paul SM (1982) Effects of tetrazole derivatives on [$^3$H]diazepam binding in vitro: correlation with convulsant potency. Eur J Pharmacol 78:353-356.

Rocha L, Ondarza-Rovira R (1999) Characterization of benzodiazepine receptor binding following kainic acid administration: an autoradiography study in rats. Neurosci Lett. 262(3):211-4.

Rudolph U, Crestani F, Benke D, Brunig I, Benson JA, Fritschy JM, Martin JR, Bluethmann H, Mohler H (1999) Benzodiazepine actions mediated by specific gamma-aminobutyric acid(A) receptor subtypes. Nature 401:796-800.

Scharfman HE, Gray WP (2007) Relevance of seizure-induced neurogenesis in animal models of epilepsy to the etiology of temporal lobe epilepsy. Epilepsia 48 Suppl 2:33-41.

Schousboe A, Waagepetersen HS (2006) Glial modulation of GABAergic and glutamatergic neurotransmission. Curr Top Med Chem 6:929-934.

Squires RF, Saederup E, Crawley JN, Skolnick P, Paul SM (1984) Convulsant potencies of tetrazoles are highly correlated with actions on GABA/benzodiazepine/picrotoxin receptor complexes in brain. Life Sci 35:1439-1444.

Tchekalarova J, Sotiriou E, Georgiev V, Kostopoulos G, Angelatou F (2005) Up-regulation of adenosine $A_1$ receptor binding in pentylenetetrazol kindling in mice: effects of angiotensin IV. Brain Res 1032:94-103.

Walsh LA, Li M, Zhao TJ, Chiu TH, Rosenberg HC (1999) Acute pentylenetetrazol injection reduces rat $GABA_A$ receptor mRNA levels and GABA stimulation of benzodiazepine binding with no effect on benzodiazepine binding site density. J Pharmacol Exp Ther 289:1626-1633.

Zgodzinski W, Rubaj A, Kleinrok Z, Sieklucka-Dziuba M (2001) Effect of adenosine $A_1$ and $A_2$ receptor stimulation on hypoxia-induced convulsions in adult mice. Pol J Pharmacol 53:83-92.

Zilles K (1985) The cortex of the rat – a stereotaxic atlas, Berlin Heidelberg: Springer

Zilles K, Palomero-Gallagher N, Schleicher A (2004) Transmitter receptors and functional anatomy of the cerebral cortex. J Anat 205(6):417-432

Zilles K, Qu MS, Kohling R, Speckmann EJ (1999) Ionotropic glutamate and GABA receptors in human epileptic neocortical tissue: quantitative in vitro receptor autoradiography. Neuroscience 94:1051-1061.

Zilles K, Schleicher A (1991) Quantitative receptor autoradiography and image analysis. Bull Assoc Anat (Nancy ) 75:117-121.

Zilles K, Schleicher A (1995) Correlative imaging of transmitter receptor distributions in human cortex. In: Autoradiography and Correlative Imaging (Stumpf WE, Solomon HF, eds), pp 277-307. San Diego: Academic Press.

Zilles K, Schleicher A, Palomero-Gallagher N, Amunts K (2002) Quantitative analysis of cyto- and receptor architecture of the human brain. In: Brain Mapping. The Methods (Toga AW, Mazziotta JC, eds), pp 573-602. Amsterdam: Elsevier.

# Fast, quantitative in situ hybridization of rare mRNAs using $^{14}$C standards and phosphorus imaging

Christian M Cremer; Markus Cremer; Jennifer Lopez Escobar; Erwin-J. Speckmann and Karl Zilles

Here, we describe a method for fast, quantitative in situ hybridization (qISH) of mRNAs using $^{33}$P-labelled oligonucleotides together with $^{14}$C-polymer standards (Microscales, Amersham Biosciences) and a phosphorus imaging system (BAS 5000 BioImage Analyzer, Raytest-Fuji). It enables a complete analysis of rare mRNAs by ISH.

We used this approach as an example for applications to quantify the expression of GluR1 and GluR2 subunit mRNAs of the α-amino-3-hydroxy-5-methyl-4-isoxazolepropionic acid (AMPA) receptor in the hippocampus of untreated rats, and after intraperitoneal application of the organo-arsenic compound dimethyl arsenic acid.

Contributions on experimental design, realization and publication:
Experimental design and practical realization were accomplished and attended by C. Cremer. The manuscript including all figures and tables was prepared by C. Cremer and subsequently reviewed, amended and approved by all co-authors.

Approximated total share of contribution in per cent: 80%

State of publication: published. J Neurosci Methods. 2009; 185(1):56-61.

## Fast, quantitative in situ hybridization of rare mRNAs using $^{14}C$ standards and phosphorus imaging

Christian M Cremer[a,b]; Markus Cremer[a]; Jennifer Lopez Escobar[a]; Erwin-J. Speckmann[c] and Karl Zilles[a,b,d]

[a]*Institute for Neuroscience and Medicine (INM-2), Research Center Jülich, D-52425 Jülich, Germany*

[b]*C. & O. Vogt Institute for Brain Research, Heinrich-Heine-University Düsseldorf, Universitätsstr. 1, D-40225 Düsseldorf, Germany*

[c]*Institute of Physiology I, Westfälische-Wilhelms-University, Robert-Koch-Str. 27a, D-48149 Münster, Germany*

[d]*JARA, Jülich Aachen Research Alliance, Research Center Jülich, D-52425 Jülich, Germany*

Corresponding author:

*Christian Cremer*

*Institute for Neuroscience and Medicine (INM-2)*

*Research Center Jülich, D-52425 Jülich, Germany*
Phone: +49-2461-61-8615 Fax: +49-2461-61-2820

Email: c.cremer@fz-juelich.de

Number of figures  6

Number of tables  1

**Abstract**

The use of radiolabelled probes for in situ hybridization (ISH) bears the advantage of high sensitivity and quantifiability. The crucial disadvantages are laborious hybridization protocols, exposition of hybridized sections to film for up to several weeks and the time consuming need to prepare tissue standards with relatively short-lived isotopes like $^{33}$P or $^{35}$S for each experiment. The quantification of rare mRNAs like those encoding for subunits of neurotransmitter receptors is therefore a challenge in ISH.

Here, we describe a method for fast, quantitative in situ hybridization (qISH) of mRNAs using $^{33}$P-labelled oligonucleotides together with $^{14}$C-polymer standards (Microscales, Amersham Biosciences) and a phosphorus imaging system (BAS 5000 BioImage Analyzer, Raytest-Fuji). It enables a complete analysis of rare mRNAs by ISH. The preparation of short-lived $^{33}$P-standards for each experiment was replaced by co-exposition and calibration of long-lived $^{14}$C-standards together with $^{33}$P-labelled brain paste standards. The use of a phosphorus imaging system allowed a reduction of exposition time following hybridization from several weeks to a few hours or days.

We used this approach as an example for applications to quantify the expression of GluR1 and GluR2 subunit mRNAs of the α-amino-3-hydroxy-5-methyl-4-isoxazolepropionic acid (AMPA) receptor in the hippocampus of untreated rats, and after intraperitoneal application of the organo-arsenic compound dimethyl arsenic acid.

**Key words**

quantitative in situ hybridization, standardization, microscales, phosphorus imaging, autoradiography, neurotransmitter receptors, dimethyl arsenic acid

## Introduction

Quantitative in situ hybridization (qISH) of mRNAs using radiolabelled complementary nucleic acid probes is a powerful method to visualize and quantify gene expression. The measurement of mRNAs using $^{33}$P- or $^{35}$S-labelled oligodeoxynucleotides is a highly sensitive, reliable technique and equivalent to RT-PCR based approaches (Broide et al., 2004). However, besides quantifiability ISH bears the advantage of visualizing mRNA expression within the native tissue providing a more discrete, anatomical analysis.

Assessment of gene expression using qISH demands some basic premises. First, a reliable standardization is needed in order to compare the results of different experiments or the expression patterns of different mRNAs. Second, a highly sensitive approach is necessary to achieve reliable results when rare mRNAs are quantified. Third, unspecific binding has to be accurately monitored to correct quantitative results from densitometric measurements. Although qISH is widely accepted as an appropriate method to analyse gene expression patterns in the brain, not all current protocols are applicable concerning standardization, sensitivity and signal to noise ratio. Protocols matching all premises for accurate qISH are often laborious and time demanding (Vizi et al., 2001). When using nucleic acid probes, deoxyribonucleotides are less delicate during handling and storage than riboprobes, since they are insensitive to degradation by ribonucleases. Nucleic acid probes for qISH are typically marked using either $^{35}$S- or $^{33}$P-labelled nucleotides. The use of $^{33}$P bears higher sensitivity than approaches using $^{35}$S due to its higher emission energy (0.25 MeV vs 0.17 MeV), thus, being preferable in qISH when measuring rare mRNAs.

Polymer standards labelled with $^{14}$C can be used for calibration of autoradiographs prepared with $^{35}$S or $^{33}$P-labelled probes (Miller, 1991; Baskin and Stahl, 1993). The use of $^{14}$C is beneficial because of its half-life (5730 years) compared to short-lived $^{33}$P (25.4 days), which spares the tedious preparation of tissue standards for each experiment.

Phosphor imaging can be used as an alternative to conventional film autoradiography reducing the time required for probe exposition several fold (Ito et al., 1995). Further advantages of phosphor imaging are its high sensitivity, the linearity of response and its high dynamic range. It is applicable for different isotopes typically used in autoradiography and we have positive experience using it with $^{18}$F as well as $^3$H (e.g., Bauer et al., 2003; Langen et al., 2007). Here, we introduce this technique in $^{33}$P qISH.

Since the organo-arsenic compound dimethyl arsenic acid (DMA$^{III}$) has been proposed to reduce the number of AMPA receptors and its subunit expression (Krüger et al., 2006; 2007), we tested our method by quantifying the expression of GluR1 and GluR2 subunit mRNAs in the hippocampus of rats which were either treated with DMA$^{III}$ or saline only.

Publications

**Materials and methods**

Animals

All experiments were performed according to the German animal welfare act and approved by the responsible governmental agency. Wistar rats (Harlan Winkelmann, Germany) were kept under standard laboratory conditions with access to food and water ad libitum. Male rats were treated at postnatal day 7 (P 0 being the day of birth) with a single intraperitoneal dose of either physiological saline containing the organo-arsenic compound dimethyl arsenic acid (670 µg $DMA^{III}$/kg bodyweight) or vehicle (saline) only.

Tissue processing

At postnatal day 35 rats were anesthetized by $CO_2$ and decapitated. The brains were removed and either used for standard calibration or were quickly frozen in -50°C isopentane and stored at -80°C. For hybridization, coronal, 20 µm thick cryostat (CM 3050; Leica) sections were serially cut at -20°C, mounted on silan coated slides and dried on a heating plate at 37°C for 15 minutes. Sections were subsequently fixed in 4% paraformaldehyde dissolved in 0.1 M phosphate buffered saline (PBS) pH 7.4 for 15 minutes, rinsed three times in PBS for 5 minutes and dehydrated in 70 % , 95 % and 100% isopropanol. Sections were stored in 100% isopropanol at 4°C until use.

Calibration of $^{14}C$-microscales using $^{33}P$-brain paste standards

Three brains were dispersed at 20,000 rpm for 5 minutes with a homogenizer (Miccra D-13, Art-Labortechnik, Germany). An aliquot was used to measure the protein content of the homogenate using Bradford reagent (Sigma-Aldrich). Further, to each 1 g of homogenate 0.6-12.5 kBq of $^{33}P$-dATP (111 TBq/mmol, PerkinElmer) were added, again homogenized for 1 minute, centrifuged for 10 minutes at 1,000 g and finally stored at -80°C over night. Of each tissue-standard three portions were weighted on an analytical balance and concentration of radioactivity (Bq/mg tissue) was estimated by liquid scintillation (Tri-Carb 2100TR, Packard BioScience). Sections of 10, 20, 30 and 50 µm thickness were prepared of each brain paste standard on a cryostat, mounted on silan coated slides, air dried and exposed to phosphor imaging plates (IPs) (BAS-SR 2025; Raytest-Fuji) together with $^{14}C$-plastic standards (Microscales; Amersham Biosciences) for 16, 48 and 72 hours.

## Oligonucleotide probes

Oligodeoxyribonucleotides (Sigma-Aldrich) complementary to rat GluR1 or GluR2 subunit mRNAs of the AMPA receptor were used for hybridization. The GluR1 probe was complementary to bases 1893-1937 of the mRNA (GenBank no. X17184.1), while the GluR2 probe was complementary to bases 2041-2076 of the respective mRNA (GenBank no. M85035.1).

## Radiolabelling

Oligonucleotide probes were 3'-labelled using $^{33}$P-dATP (111 TBq/mmol, PerkinElmer) and terminal deoxynucleotidyl transferase (TdT). 0.5 pmol probe, 1 MBq of $^{33}$P-dATP and 30 units of recombinant TdT (Terminal Deoxynucleotidyl Transferase, Recombinant, Promega) were incubated at 37°C for 90 minutes in a final volume of 10 µl 100 mM potassium cacodylate (pH 6.8 at 25°C), 0.1 mM dithiothreitol and 1 mM $CoCl_2$. The reaction buffer was provided by the supplier of TdT. To terminate the reaction, the assay was chilled on ice for 1 minute, diluted with 40 µl of STE-buffer (0.9 % NaCl, 10 mM Tris-HCl, 1 mM EDTA, pH 8.0) and the probe was subsequently purified from unincorporated nucleotides using sephadex columns (ProbeQuant G-50, GE Healthcare) according to the manufacturers' instructions. Finally, the purified probe was dissolved in bi-distilled water to a final volume of 150 µl and specific activity (SA) was estimated by triplicated liquid scintillation measurement. A typical probe exhibited a SA between 300-600 TBq/mmol.

## Hybridization

Sections were air dried and pretreated for 15 minutes in 1x standard saline citrate (SSC, pH 7.0) at room temperature containing 0.1 mg/ml proteinase K (Sigma-Aldrich). Subsequently, sections were washed two times in 1x SSC for 5 minutes. Labelled oligonucleotides were diluted (30 pM) in hybridization buffer containing 50% de-ionized formamide, 4x SSC pH 7.0, 1 mM sodium pyrophosphate, 0.25 mg/ml hydrolysed salmon sperm DNA, 0.1 mg/ml polyadenylic acid and 100 mg/ml dextran sulfate. 50 µl of probe buffer were added to each section, covered with Parafilm™ and incubated in a humid chamber over night at 42°C. Unspecific binding was monitored in presence of 100-fold excess of unlabelled oligonucleotide. Following incubation, sections were rinsed for 10 minutes in 1x SSC at room temperature, 20 minutes in 1x SSC at 60°C, 5 minutes in distilled water at room temperature, dehydrated in 70%, 95% and 100% isopropanol and finally air dried. Sections were exposed for 48 h to IPs together with $^{14}$C-microscales.

Image analysis

A BAS 5000 BioImage Analyzer (Raytest-Fuji) was used to scan the IPs and measure photo stimulated luminescence (PSL), resulting in digital autoradiographic images.

Standard image analysis software (AIDA 2.31; Raytest) was used for image analysis. Densitometric measurement of mRNA expression was performed for each receptor subunit on 3-5 serial sections of each six brains per group. Specific binding was defined as total binding minus unspecific binding and mean densities were calculated for regions of interest (ROIs), i.e. hippocampal subfields CA1, CA2/3 and dentate gyrus (DG). ROIs were traced according to the cytoarchitectonic atlas of Zilles (1985).

The measurements of optical density (PSL/mm$^2$) within ROIs were first converted into Bq/mg tissue using a calibration curve based on the values of $^{14}$C-microscales coexposed with $^{33}$P tissue standards. In a second step estimation of SA of the labelled probe and the protein content of tissue standards were used to calculate mRNA copy numbers per mg protein (fmol x $10^{-3}$/mg protein).

Data analysis

Data are expressed as mean ± SD unless otherwise indicated. Results of qISH experiments have been statistically analysed using one-way analysis of variance with repeated measurements.

## Results

Influence of section thickness and exposure time

Brain paste sections (10-150 µm) were exposed to IPs for 16 h, to evaluate the influence of section thickness on signal intensity. Three sections were used for each nominal thickness. The increase of PSL/mm$^2$ with section thickness was fitted by second order polynomial regression (r = 0.996; Figure 1). The increment was approximately linear between 10 and 60 µm (r = 0.980). Section thickness for standard calibration and qISH was kept within this range.

To verify that the amount of radioactivity in the brain paste section was linearly increasing with section thickness (Bq/µm thickness), each three sections were additionally measured by liquid szintillation (Figure 1, inlay). The increase of radioactivity proved to be linear within this range (r = 0.997).

To investigate the influence of exposure time on PSL/mm$^2$ $^{14}$C-plastic standards were exposed to IPs for 4-120 hours and the PSL/mm$^2$ measured for one level (3.68 Bq/mg polymer). The PSL/mm$^2$ was plotted against exposure time and fitted by second order polynomial regression (r = 0.997, Figure 2). The increase of PSL/mm$^2$ with exposure time was approximately linear between 4-48 hours (r = 0.994).

Calibration of $^{14}$C-plastic standards

To calibrate $^{14}$C-platic standards for qISH $^{33}$P-labelled brain homogenate sections (10, 20, 30, 50 µm thick) were exposed with $^{14}$C-plastic standards against IPs for 16, 48 and 72 h. For each section thickness and exposure period the PSL/mm$^2$ was plotted against radioactivity concentration (Bq/mg tissue) and the relationship between PSL/mm$^2$ and Bq/mg was estimated by linear regression (Figure 3). The underlying equations were used to derive tissue-equivalent radioactivity from PSL/mm$^2$ produced by the different levels of $^{14}$C-standards and the mean ± standard deviation was calculated (Table 1).

Using the calculations of protein content from brain homogenates (54.3 ± 6.5 mg protein/g tissue), measurements of tissue-equivalent radioactivity can be expressed as Bq/mg protein.

## GluR1 and GluR2 mRNA expression

Following calibration, $^{14}$C-microscales were used as standards to measure the mRNA levels of GluR1 and GluR2 mRNAs (Bq/mg protein) in the hippocampal subfields CA1, CA2/3 and DG of rats, treated with either a single dose of DMA$^{III}$ or saline only. The SA (Bq/fmol) of the respective oligonucleotide was then used to calculate the regional mRNA density given as fmol x $10^{-3}$/mg protein.

Treatment with DMA$^{III}$ significantly reduced the mRNA levels of both GluR1 and GluR2 subunits in the ROIs investigated (Figures 4 and 5). While the levels of GluR1 mRNA of DMA$^{III}$-treated rats proved to be significantly reduced in all ROIs (-48-61%, $p < 0.001$ or $0.01$), GluR2 mRNA density was only found to be significantly changed in CA1 (-35%, $p < 0.05$) and CA2/3 (-58%, $p < 0.01$).

# Quantitative in situ hybridization

**Figure 1**: Influence of $^{33}$P brain paste section thickness on phosphorus imaging signal intensity. The increase of PSL/mm$^2$ with section thickness was fitted by second order polynomial regression. The increment was approximately linear between 10 and 60 µm (dashed line, r = 0.980). Inlay: Radioactive content measured by liquid scintillation increases linearly with section thickness, demonstrating homogenous distribution of $^{33}$P within the tissue standard.

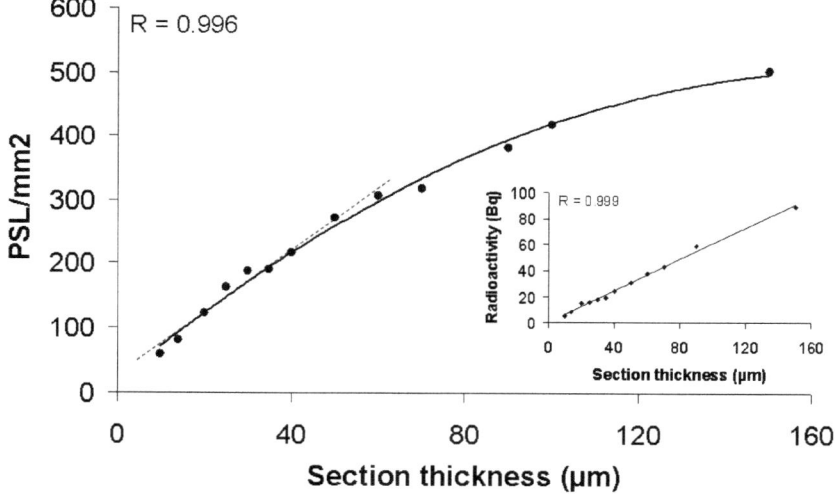

**Figure 2**: Influence of exposure time on signal intensity of $^{14}$C-polymer standards exposed against ß-sensitive phosphorus imaging plates. A radioactive $^{14}$C standard (3.68 Bq/mg polymer) was exposed against IPs for 4-120 hours and the PSL/mm$^2$ was measured. A curve was fitted by second order polynomial regression. The increment of signal intensity with exposure time was approximately linear between 4-48 hours (dashed line, r = 0.994)

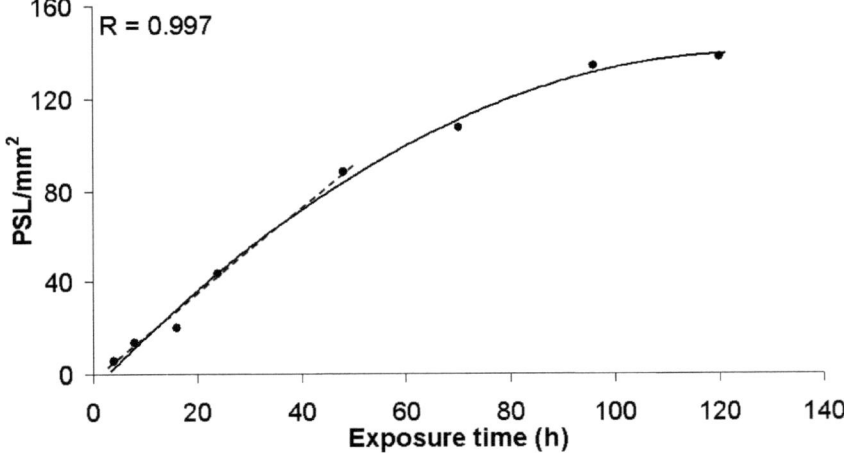

# Quantitative in situ hybridization

**Figure 3**: Calibration of $^{14}$C-plastic standards for qISH using $^{33}$P-labelled brain homogenate sections. Brain paste standards with increasing radioactive concentrations (0.6-73.8 Bq/mg tissue) were used. Sections of 10, 20, 30 and 50 µm thickness were exposed against ß-sensitive IPs for 16-72 hours together with $^{14}$C-standards to derive tissue-equivalent radioactivity for the different levels of $^{14}$C (for details see text).

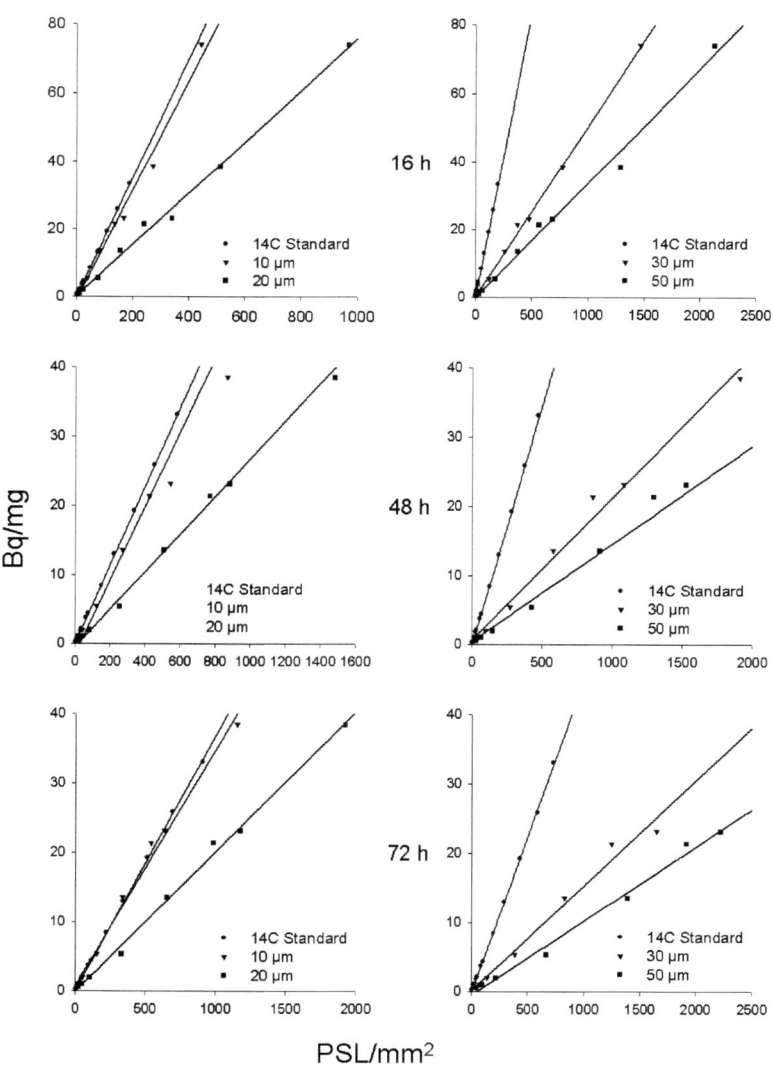

**Table 1**: Calculation of tissue-equivalent radioactivity of $^{14}$C polymer standards using the equations of linear regression derived from co-exposition with $^{33}$P brain paste standards.

| $^{14}$C Standard (Bq/mg) | Tissue-equivalent radioactivity ($^{33}$P Bq/mg) | | | |
|---|---|---|---|---|
| | 10 μm | 20 μm | 30 μm | 50 μm |
| 0.19 | 0.17 ± 0.01 | 0.09 ± 0.00 | 0.06 ± 0.01 | 0.04 ± 0.01 |
| 0.37 | 0.34 ± 0.02 | 0.18 ± 0.02 | 0.13 ± 0.01 | 0.08 ± 0.00 |
| 0.93 | 0.84 ± 0.05 | 0.44 ± 0.03 | 0.31 ± 0.01 | 0.21 ± 0.01 |
| 1.10 | 1.04 ± 0.03 | 0.55 ± 0.06 | 0.39 ± 0.04 | 0.26 ± 0.03 |
| 1.82 | 1.70 ± 0.08 | 0.89 ± 0.05 | 0.62 ± 0.03 | 0.42 ± 0.02 |
| 2.10 | 2.00 ± 0.01 | 1.06 ± 0.10 | 0.72 ± 0.06 | 0.49 ± 0.05 |
| 3.68 | 3.34 ± 0.06 | 1.76 ± 0.13 | 1.23 ± 0.05 | 0.83 ± 0.03 |
| 4.40 | 4.00 ± 0.05 | 2.11 ± 0.20 | 1.46 ± 0.13 | 0.99 ± 0.09 |
| 8.40 | 7.70 ± 0.06 | 4.06 ± 0.38 | 2.76 ± 0.20 | 1.87 ± 0.15 |
| 13.00 | 11.84 ± 0.15 | 6.25 ± 0.62 | 4.16 ± 0.30 | 2.82 ± 0.22 |
| 19.20 | 17.59 ± 0.15 | 9.29 ± 0.96 | 6.13 ± 0.41 | 4.15 ± 0.30 |
| 25.80 | 23.59 ± 0.42 | 12.46 ± 1.40 | 8.30 ± 0.60 | 5.62 ± 0.44 |
| 33.10 | 30.51 ± 0.79 | 16.13 ± 1.94 | 10.37 ± 0.65 | 7.02 ± 0.47 |

**Figure 4**: Quantification of GluR1 and GluR2 mRNA expression levels in the hippocampus of rats, treated with a single intraperitoneal injection of DMA$^{III}$ or vehicle. Treatment with DMA$^{III}$ leads to significant reduction of both GluR1 and GluR2 mRNAs in the hippocampal subfields. Mean + SEM; * p < 0.05; ** p < 0.01; *** p < 0.001.

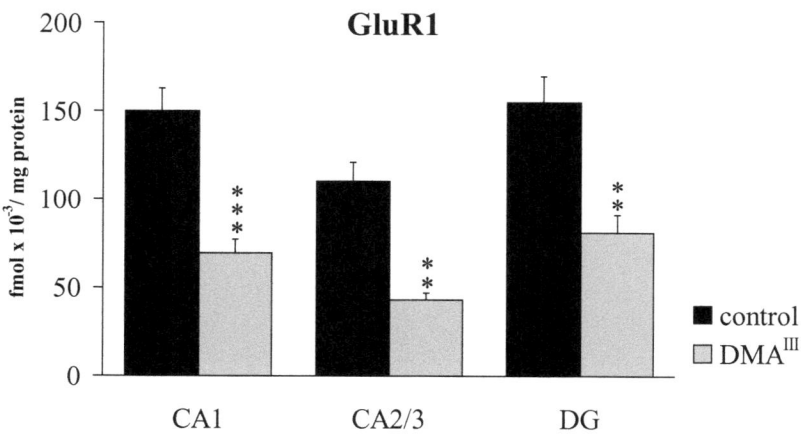

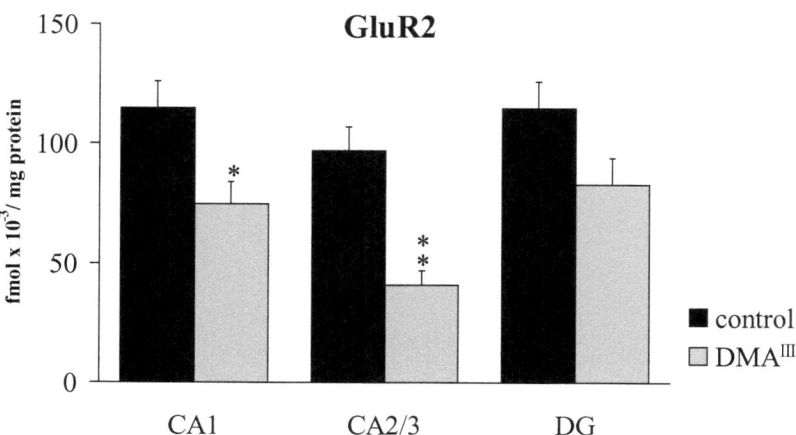

Publications

**Figure 5**: Intraperitoneal injection of DMA$^{III}$ reduces the expression of GluR1 and GluR2 subunit mRNAs in the rat hippocampus. Contrast enhanced and colour coded autoradiographs of quantitative in situ hybridization visualize the distribution and relative expression levels of the respective mRNA. For each subunit autoradiographs have been colour coded using the same scaling.

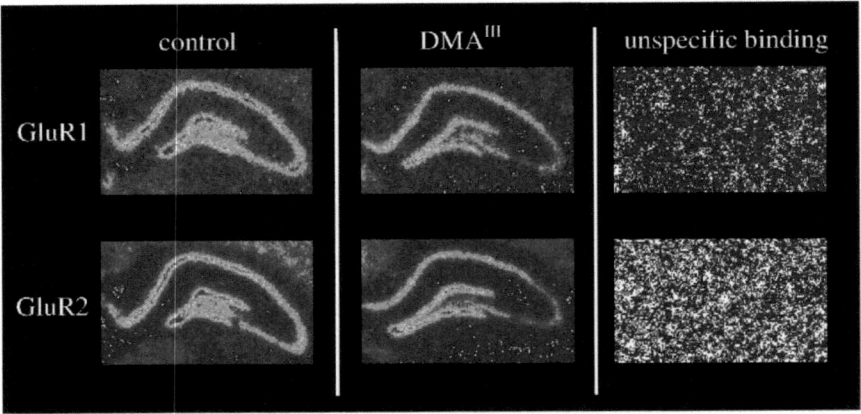

## Discussion

The aim of the present study was to establish a fast method for analysing the expression levels of neurotransmitter receptor mRNAs in the brain. Our results demonstrate that mRNAs of neurotransmitter receptor subunits can be accurately measured using the method described.

Using in vitro receptor autoradiography, a maximum image resolution of 10 µm can be achieved with $^3$H-labelled ligands and conventional film (Zilles et al., 2002). In comparison, IPs exhibit only a limited resolution of 50 µm, thus, resulting in a disadvantage concerning image quality of phosphorus imaging when working with $^3$H-labelled probes. However, the use of $^3$H-labelling is generally regarded as inappropriate for qISH, since quenching of $^3$H radiation by lipophilic structures and proteins within the tissue would adulterate quantitative analysis. Thus, probes for qISH are typically labelled using $^{35}$S or $^{33}$P, which are less affected by tissue-inherent quenching due to their higher emission energies. Here, the maximum spatial resolution within a $^{33}$P-labelled tissue sample was shown to be within the range of 140-160 µm (Figure 1). Conclusively, when working with $^{33}$P, there is no advantage in image resolution using conventional film. The use of phosphorus imaging is a suitable approach to generate autoradiographs in qISH within only a few hours or days. Depending on specific activity of the hybridization probe, images of high quality can be produced in less than 8 h. However, we generally prefer exposition times between 16-48 hours for optimal results rich in contrast. Exposition for more than 72 hours does not substantially improve signal intensity or image quality respectively (cf. Fig. 2).

In order to simplify the experimental workflow we reduced our protocol to a few necessary steps. However, although hybridization experiments worked without any pretreatment of sections, we observed noteworthy enhancement of ISH signals when applying proteinase K treatment to increase probe penetration. Performance of ISH for low abundant mRNAs without this step did often result in poor signal to noise ratio. Although often suggested, we did not observe an improvement in signal to noise ratio when adding prehybridization steps or further high stringency washes following hybridization. The time for preparing the sections can be further short cut by immediate use of sections for hybridization without tissue fixation. Although this approach provides feasible hybridization results, the histological quality of sections was drastically reduced by omitting the fixation. Furthermore, when fixed tissue is used for ISH the degradation of target RNA by ribonucleases is a negligible risk (Tongiorgi et al., 1998).

To test our approach, we measured the expression of GluR1 and GluR2 subunit mRNAs of the AMPA receptor in the hippocampus of saline treated rats and after a single application of DMA$^{III}$. Our results showed that DMA$^{III}$ significantly reduces the mRNA expression levels of GluR1 and GluR2 subunits in the hippocampus. Thus, mRNAs of neurotransmitter receptor subunits show the awaited reduction after using the present method.

It has to be noted though that quantitative ISH does not reflect absolute in vivo quantities of mRNA copies, since factors affecting ISH are numerous (e.g. tissue fixation, pH, proteinase treatment, probe length and penetration, hybridization temperature). Although attempts have been made to normalize quantitative ISH results, e.g. by calculation of maximum hybridization capacities (Vizi and Gulya, 2000), it remains difficult to compare quantitative data gained by different protocols or experimental approaches. Therefore, we preferred a quantitative ISH protocol which is a good compromise between technical simplicity, standardization and reliability that can be used in large scale experiments.

In conclusion, we here described a protocol for quantitative ISH that can easily be performed within two or three days. Additionally, our approach does neither require particular precautions against probe degradation. Thus, the method can be easily used in any laboratory to quantify the expression of high and low abundant mRNAs.

## Acknowledgements:

The authors appreciate the excellent technical assistance of S. Buller.

## References

Baskin DG, Stahl WL. Fundamentals of quantitative autoradiography by computer densitometry for in situ hybridization, with emphasis on 33P. J Histochem Cytochem. 1993;41(12):1767-76.

Bauer A, Holschbach MH, Cremer M, Weber S, Boy C, Shah NJ, Olsson RA, Halling H, Coenen HH, Zilles K. Evaluation of 18F-CPFPX, a novel adenosine A1 receptor ligand: in vitro autoradiography and high-resolution small animal PET. J Nucl Med. 2003;44(10):1682-9.

Broide RS, Trembleau A, Ellison JA, Cooper J, Lo D, Young WG, Morrison JH, Bloom FE. Standardized quantitative in situ hybridization using radioactive oligonucleotide probes for detecting relative levels of mRNA transcripts verified by real-time PCR. Brain Res. 2004;1000(1-2):211-22.

Ito T, Suzuki T, Lim DK, Wellman SE, Ho IK. A novel quantitative receptor autoradiography and in situ hybridization histochemistry technique using storage phosphor screen imaging. J Neurosci Methods. 1995;59(2):265-71.

Krüger K, Gruner J, Madeja M, Hartmann LM, Hirner AV, Binding N, Mushoff U. Blockade and enhancement of glutamate receptor responses in Xenopus oocytes by methylated arsenicals. Arch Toxicol. 2006;80:492-501.

Krüger K, Repges H, Hippler J, Hartmann LM, Hirner AV, Straub H, Binding N, Mushoff U. Effects of dimethylarsinic and dimethylarsinous acid on evoked synaptic potentials in hippocampal slices of young and adult rats. Toxicol Applied Pharmacol. 2007;225:40-46.

Langen KJ, Salber D, Hamacher K, Stoffels G, Reifenberger G, Pauleit D, Coenen HH, Zilles K. Detection of secondary thalamic degeneration after cortical infarction using cis-4- 18F-fluoro-D-proline. J Nucl Med. 2007;48(9):1482-91.

Miller JA. The calibration of 35S or 32P with 14C-labeled brain paste or 14C-plastic standards for quantitative autoradiography using LKB Ultrofilm or Amersham Hyperfilm. Neurosci Lett. 1991;121(1-2):211-4.

Publications

Tongiorgi E, Righi M, Cattaneo A. A non-radioactive in situ hybridization method that does not require RNAse-free conditions. J Neurosci Methods. 1998;85(2):129-39.

Vizi S, Gulya K. Calculation of maximal hybridization capacity (Hmax) for quantitative in situ hybridization: a case study for multiple calmodulin mRNAs. J Histochem Cytochem. 2000;48(7):893-904.

Vizi S, Palfi A, Hatvani L, Gulya K. Methods for quantification of in situ hybridization signals obtained by film autoradiography and phosphorimaging applied for estimation of regional levels of calmodulin mRNA classes in the rat brain. Brain Res Brain Res Protoc. 2001;8(1):32-44.

Zilles K. The Cortex of the Rat – A Stereotaxic Atlas. Springer, Berlin, 1985.

Zilles K, Schleicher A, Palomero-Gallagher N, Amunts K. Quantitative analysis of cyto- and receptor architecture of the human brain. In: Brain Mapping. The Methods (Toga AW, Mazziotta JC, Eds);573-602. Elsevier, Amsterdam, 2002.

**Inhibition of glutamate/glutamine cycle in vivo results in decreased benzodiazepine binding and differentially regulated GABAergic subunit expression in the rat brain**

Christian M Cremer; Hans-Jürgen Bidmon; Boris Görg; Nicola Palomero-Gallagher; Jennifer Lopez Escobar; Erwin-J. Speckmann, and Karl Zilles

The astrocytic enzyme glutamine synthetase (GS) is a key regulator of glutamate and GABA metabolism in the glutamate/glutamine cycle (GGC). Inhibition of GS results in changes of neurotransmitter release and recycling. However, little is known about the influence of GGC on neurotransmitter receptor expression. In the pentylenetetrazole model of epilepsy GS becomes nitrated and partially inhibited, and we demonstrated alterations of neurotransmitter receptor expression in the same model. Thus, we hypothesized similar changes of neurotransmitter receptor expression when GS is inhibited in vivo.

Based on our findings, we conclude that the glutamate/glutamine cycle directly influences GABAergic neurotransmission by regulating $GABA_A$ subunit composition, thus affecting its modulation by endogenous benzodiazepines.

Contributions on experimental design, realization and publication:
The autoradiographic experiments, saturation analysis and in situ hybridizations as well as the analysis of the respective results were performed by C. Cremer. All technical assistance is acknowledged in the manuscript. The manuscript including all figures and tables was prepared by C. Cremer and subsequently reviewed, amended and approved by all co-authors.

Approximated total share of contribution in per cent: 70%

Current state of publication: in press, Epilepsia

Publications

**Inhibition of glutamate/glutamine cycle in vivo results in decreased benzodiazepine binding and differentially regulated GABAergic subunit expression in the rat brain**

Running title: Glutamate/glutamine cycle and $GABA_A$ subunits

Cremer, Christian M[1,2]; Bidmon Hans-Jürgen[2]; Görg, Boris[3]; Palomero-Gallagher[1,2], Nicola; Lopez Escobar, Jennifer[1], Speckmann, Erwin-J[4], and Zilles, Karl[1,2]

[1]Institute of Neuroscience and Medicine (INM-2), Research Center Jülich, Jülich, Germany
[2]C & O Vogt Institute for Brain Research, Heinrich-Heine-University, Düsseldorf, Germany,
[3]Clinic for Gastroenterology, Hepatology and Infectiology, Heinrich-Heine-University, Düsseldorf, Germany
[4]Institute of Physiology I, Westfälische-Wilhelms-University, Münster, Germany.

Corresponding author:
Christian Cremer, Institute of Neuroscience and Medicine (INM-2), Research Center Jülich, D-52425 Jülich, Germany
email: c.cremer@fz-juelich.de
tel: +49-2461-61-8615
fax: +49-2461-61-2820

Number of figures: 7
Number of tables: 1

Number of words (4497):
Summary:    250
Introduction: 495
Methods:    1379
Results:    688
Discussion: 1744

## Summary

**Purpose:** The astrocytic enzyme glutamine synthetase (GS) is a key regulator of glutamate and GABA metabolism in the glutamate/glutamine cycle (GGC). Inhibition of GS results in changes of neurotransmitter release and recycling. However, little is known about the influence of GGC on neurotransmitter receptor expression. In the pentylentetrazole model of epilepsy GS becomes nitrated and partially inhibited, and we demonstrated alterations of neurotransmitter receptor expression in the same model. Thus, we hypothesized similar changes of neurotransmitter receptor expression when GS is inhibited in vivo.

**Methods:** Rats were treated with a single dose (100 mg/kg bodyweight) of L-methionine sulfoximine (MSO), an irreversible inhibitor of GS. We used $^3$H-receptor autoradiography to measure glutamatergic (α-amino-3-hydroxy-5-methyl-4-isoxazol-propionic acid (AMPA), kainate, NMDA), GABAergic (GABA$_A$, GABA$_B$ and GABA$_A$ associated benzodiazepine (BZ) binding sites), dopamine D$_1$ and adenosine A$_1$ receptor subtypes. In addition, we performed saturation analysis of BZ binding sites on cerebral membrane homogenates and investigated the expression of GABA$_A$ α$_1$ and γ$_2$ subunits (which primarily mediate BZ binding) by western blot analysis.

**Results:** We demonstrated a significant reduction of BZ binding in the somatosensory, piriform and entorhinal cortices and in the amygdala 24h and 72h after MSO-treatment. Saturation analysis revealed decreased BZ-binding (B$_{max}$) on cerebral membrane homogenates 72h after MSO-treatment without changes in binding site affinity (K$_D$). Furthermore, we found differential changes of α$_1$, γ$_2$ and phosphorylated γ$_2$ subunits following MSO treatment.

**Conclusion:** Based on our findings, we conclude that the glutamate/glutamine cycle directly influences GABAergic neurotransmission by regulating GABA$_A$ subunit composition, thus affecting its modulation by endogenous benzodiazepines.

**Keywords:** L-methionine sulfoximine, neurotransmitter receptors, GABA$_A$, benzodiazepine binding, autoradiography, in situ hybridization

## Introduction

The glutamate/glutamine cycle (GGC) is a major regulator of glutamate and GABA metabolism (reviewed by Bak et al., 2006). Glutamate is released into the synaptic cleft during neurotransmission and subsequently removed via uptake into astrocytes. The astrocytic enzyme glutamine synthetase (GS) catalyzes the generation of glutamine from glutamate (Martinez-Hernandez et al., 1977), which is transferred via system N and A transporters into neurons (Chaudhry et al., 2002) where it is deaminated by phosphate-activated glutaminase, thus producing glutamate (Kvamme et al., 2000). The cycle is completed by transport of glutamate into vesicles. Although GABAergic interneurons are capable of recycling GABA by specific transporters, a major source of vesicular GABA is derived from decarboxylation of glutamate (Martin and Tobin, 2000).

L-methionine-sulfoximine (MSO) irreversibly inhibits GS in vivo and in vitro (Lamar and Sellinger, 1965; Ronzio et al., 1969). A seizure model was recently established in rats using chronic hippocampal microinfusion of MSO (Eid et al., 2008) which induces recurrent epileptic activity, thus modelling the pathology of mesial temporal lobe epilepsy (Eid et al., 2004) and implying GS is a putative target for pharmacological intervention.

Neurotransmitter alterations following GS-inhibition have numerously been reported for different brain regions and animal models. Application of MSO in vivo transiently decreases whole brain GABA content (Stransky, 1969). GABA, glutamate and glutamine concentrations are differentially altered in the neostriatum and globus pallidus of MSO-treated rats (Fonnum and Paulsen, 1990), while the striatal release of dopamine is affected by MSO infusion (Rothstein and Tabakoff, 1982). Furthermore, the GGC regulates synaptic GABA content under physiological conditions (Liang et al., 2006). Vice versa, activation of neurotransmitter receptors influences GGC activity, that is, an activation of glutamatergic NMDA receptors reduces GS activity, likely by increasing the formation of nitric oxide and resulting nitration of GS (reviewed by Rodrigo and Felipo, 2007). Together, these data suggest a close regulatory relation between neurotransmitters, their receptors and the GGC.

Intraperitoneal injection of MSO leads to enhanced heat shock protein 27 (HSP-27) expression in all astrocytes (Bidmon et al., 2008). In the pentylentetrazole model of epilepsy glial HSP-27 response exhibits a much more focal pattern and, furthermore, GS becomes nitrated and its activity reduced (Bidmon et al., 2005, 2008).

Additionally, we have shown that these changes largely coincide with alterations of different neurotransmitter receptor densities in a region specific manner (Cremer et al., 2009a). Therefore, we proposed that disturbances of GGC might influence the expression of neurotransmitter receptors in vivo.

We performed a multi receptor study to investigate the effects of GS inhibition on the densities, affinity and subunit composition of neurotransmitter receptors of the glutamatergic, GABAergic and dopaminergic systems. Since alterations of glutamate recycling could result in changes of astrocytic glutamate transport (Rothstein and Tabakoff, 1985), we further investigated the expression of the major glutamate transporters GLAST and Glt-1.

We show that in vivo inhibition of GS decreases BZ binding and differentially changes $GABA_A$ subunit composition, suggesting a mechanism in which the GGC is a regulatory element of GABAergic neurotransmission.

## Methods

### Animals
8-week-old, male Wistar rats (220-250 g bodyweight, Charles River, Sulzfeld, Germany) were used for all experiments and housed under standard conditions (Bidmon et al., 2005). All experiments were approved by the responsible governmental agency and conducted according to the German Animal Welfare Act.

### MSO-treatment
Rats were intraperitoneally injected with either a single dose of MSO (n = 24 animals, 100 mg/kg bodyweight, Sigma-Aldrich, Seelze, Germany) dissolved in physiological saline or vehicle (n = 12). Behavioural effects of MSO were observed as described (Rao and Murthy, 1988; Yamamoto et al., 1989). MSO treated rats behaved normal until 4h after dosing, when akinesia and ataxia progressively appeared. First signs started when the animals moved their head and thorax slowly in a pendulum like motion when trying to focus. The rats exhibited seizures approximately 8-13 hours after MSO treatment and recovered over night. MSO treated rats were divided in two groups which were sacrificed by decapitation 24h or 72h after treatment. Control animals were sacrificed 24h after the treatment. The brains were removed and frozen in isopentane at -50°C.

### GS-activity assay
GS activity was measured by γ-glutamyl transferase reaction (Webb & Brown, 1976) and has been described (Bidmon et al., 2008). Briefly, rats were decapitated, the cerebral cortex dissected and frozen on dry ice. Aliquots of cortical homogenates containing equal protein amounts in a total volume of 100μl were incubated with 900 μl of reaction mixture at 37°C. The reaction was terminated by adding 1 ml stop solution and centrifuged at 20,000 g (4°C). Absorbance was determined in protein cleared supernatants at 500 nm. GS activity of cerebellar protein homogenates of controls was set to 1 and activity found in MSO treated animals was expressed as a fraction thereof.

## ³H-receptor autoradiography

Unfixed, frozen hemispheres were serially sectioned (20 µm) in the coronal plane using a cryostat (Leica, Wetzlar, Germany) and were processed for quantitative in vitro receptor autoradiography according to standard protocols (Zilles et al., 2002; Zilles et al., 2004; Cremer et al., 2009a). Briefly, sections were washed, incubated in buffer containing the ³H-ligand and finally rinsed in buffer. Unspecific binding was monitored by incubation in presence of excess non-radioactive competitor. Sections were exposed to ß-sensitive imaging plates (BAS-TR 2025; Raytest-Fuji, Straubenhardt, Germany) for 72 hours in the presence of standards of known radioactive concentrations (GE Healthcare, Freiburg, Germany).

## Saturation analysis

The principles of saturation analysis have been described (Basile, 1997). Experiments were carried out as recently described (Palomero-Gallagher et al., 2009) with some modifications. Fresh frozen hemispheres (without cerebellum) were transferred into 10 ml of ice cold tris-citrate buffer (50 mM, pH 7.4) and homogenized for 1 minute with a homogenizer (Miccra D-13, Art-Labbortechnik, Müllheim, Germany) at 15.000 rpm. Homogenates were washed four times with 10 ml tris-citrate buffer and subsequent centrifugation (4°C, 14.800 g, 60 min) and stored at -80°C. For analysis probes were thawed, centrifuged (4°C, 14.800 g, 60 min) and the pellet diluted in incubation buffer (170 mM Tris-HCl, pH 7.4). Protein concentration of homogenates (0.3-0.6 mg/ml) was photometrically determined prior to experiments.

*Filtration assay.* Final assay volume was 300 µl. In glass reaction tubes 50 µl aliquots of ³H-Flumazenil (10 concentrations) were added to 100 µl of brain homogenate and 150 µl of incubation buffer. Triplicated measurement was performed for each concentration by scintillation spectrometry (Tri-Carb 2100, Canberra-Packard, Dreieich, Germany). Ligand concentrations in the assay ranged between 0.2-20 nM. Unspecific binding was monitored by excess addition of a specific competitor (2 µM Clonazepam, Sigma). Concentration of protein in the assay ranged between 0.1 - 0.2 mg/ml. To quantify filter binding, brain homogenate in the assay was replaced by 100 µl of buffer. After incubation binding was assessed by rapid filtration (< 5 s) over GF/B glass fiber filters, followed by 3 washes (< 10 s) with 5 ml buffer using a cell harvester (M-48, Brandel, Gaithersburg, MD, USA). Filters were transferred into vials containing 10 ml scintillation fluid (Ready Safe, Beckman Coulter, Krefeld, Germany), left shaking over night and measured by scintillation spectrometry.

Specific binding was determined as the difference between total binding and unspecific binding. Receptor affinity ($K_D$) and density ($B_{max}$) were calculated by computerized non-linear regression analysis (PRISM 2.0, GraphPad Software, San Diego, CA, USA). To monitor quality of filtration assay a ratio was calculated determining binding specificity as percent of total binding: (unspecific binding - filter binding) * 100 / (total binding – filter binding).

In situ hybridization (ISH)

Quantitative ISH was performed as recently described (Cremer et al., 2009b). Coronal 20 µm cryosections (see above) were fixed in 4% paraformaldehyd in phosphate buffered saline (PBS), pH 7.4, for 30 minutes, washed three times in PBS for 10 minutes, dehydrated in 70% and 90% ethanol and stored in 100% ethanol at 4°C.

*Oligodeoxyribonucleotides* (Sigma-Aldrich) complementary to rat $\alpha_1$ or $\gamma_2$ subunit mRNAs of the $GABA_A$ receptor were used for hybridization. The $\alpha_1$ probe was complementary to mRNA bases 67-116 (GenBank no. AY574250.1), the $\gamma_2$ probe was complementary to mRNA bases 769-808 (GenBank no. NM_183327.1).

Oligonucleotides were 3'-labelled using $^{33}$P-dATP (111 TBq/mmol, Perkin Elmer, Waltham, MA, USA) and Terminal Deoxynucleotidyl Transferase (TdT) reaction kit (Promega, Mannheim, Germany). Probes were purified using sephadex columns (ProbeQuant G-50, GE Healthcare, Munich, Germany), diluted 30 pM in buffer (50% deionized formamide, 4x standard saline citrate (SSC) pH 7.0, 1 mM sodium pyrophosphate, 0.25 mg/ml hydrolysed salmon sperm DNA, 0.1 mg/ml polyadenylic acid and 100 mg/ml dextran sulfate) and hybridization was performed over night at 42°C in humidified hybridization chambers. Unspecific binding was monitored by hybridization in presence of 100-fold excess of unlabelled oligonucleotide.

Sections were washed for 10 minutes in SSC at room temperature, 30 minutes in SSC at 60°C, followed by SSC, 0.1 x SSC, and distilled water (5 minutes each) at room temperature and dehydrated in ethanol. Finally, sections were exposed to ß-sensitive imaging plates (BAS-SR 2025; Raytest-Fuji) for 48h together with $^{14}$C-microscales (Amersham).

Image analysis
Following exposition imaging plates from $^3$H-autoradiography or ISH were scanned using a BAS-5000 reader (Raytest-Fuji) and the resulting linearized grey-scale image was used for densitometric measurement using standard image analysis software (AIDA 2.31; Raytest). Regions of interest (ROI, supplementary figure 1) were traced according to the cytoarchitectonic atlas of Zilles (1985), i.e. amygdala, CA1, dentate gyrus, substantia nigra and putamen, as well as entorhinal, somatosensory, piriform and retrosplenial granular cortices. In each animal and ROI, grey values were measured in 3-5 randomised sections. The standards were used to compute a transformation curve indicating the relationship between grey values in the autoradiographs and concentration of radioactivity in the tissue. Binding site densities were calculated as described in detail (Zilles and Schleicher, 1991, 1995; Zilles et al., 2002).

Western blot analysis
Protein was extracted from brain tissue by ultrasonification at 4°C using lysis buffer containing 20 mmol/l Tris/HCl (pH 7.4), 1% Triton X-100, 140 mmol/l NaCl, 1 mmol/l EDTA, 1 mmol/l EGTA, 10 mmol/l NaF, 10 mmol/l Na-pyrophosphate, 1 mmol/l sodium vanadate, 20 mmol/l ß-glycerophosphate and protease inhibitor mixture (Roche, Mannheim, Germany). The homogenized lysates were centrifuged at 20,000 g at 4°C. Protein concentration was estimated using Bradford reagent according to the manufacturer's protocol (BioRad, Munich, Germany). For SDS–gel electrophoresis and Western blot analysis protein extracts were added to an identical volume of 2 × gel loading buffer (pH 6.8), containing 200 mmol/L dithiothreitol. After heating to 95°C for 3 min, the samples were subjected to gel electrophoresis. Gels were then equilibrated with transfer buffer (39 mmol/L glycine, 48 mmol/L Tris–HCl, 0.03% SDS, and 20% methanol). Proteins were transferred to nitrocellulose membranes using a semi-dry transfer apparatus (Biometra, Göttingen, Germany). Membranes were blocked in 3% bovine serum albumine solubilized in 20 mmol/L Tris–HCl (pH 7.5) containing 150 mmol/L NaCl and 0.1% Tween20, and then incubated for 2h with the respective primary antibody (1:1,000; GABA$_A$-R-α1 #65269; GABA$_A$-R-γ2 #49961; GABA$_A$-R-γ2-phospho #73183 Abcam and GLAST #2064 Tocris Bioscience, Bristol, UK; GLT-1 #PC154 Calbiochem, San Diego, CA, USA) at room temperature. Following washing and incubation with horseradish peroxidase-

coupled anti-mouse-IgG or anti-rabbit antibody diluted 1:10,000 at room temperature for 2h, blots were washed again and developed using enhanced chemiluminescent detection (Amersham, Braunschweig, Germany). Densitometric analysis was performed with the Kodak Image Station 4400, using the Kodak 1D Molecular Imaging software.

Data analysis

Results from n independent experiments are expressed as mean ± SEM unless otherwise indicated. Data were statistically analysed for significant differences by two tailed Student's t-test or one way analysis of variance (ANOVA) with repeated measurements, where $p < 0.05$ was considered statistically significant.

## Results

### MSO application induces long term inhibition of glutamine synthetase activity

Intraperitoneal injection of MSO (100 mg/kg) induced a long lasting inhibition of GS-activity in cortical homogenates (Fig. 1). MSO reduced GS-activity to $30 \pm 5$ % ($p < 0.0005$) of control values measured 24h after treatment. After 72h MSO still inhibited GS, but activity recovered to $60 \pm 9$ % ($p < 0.01$) of controls. Since MSO irreversibly inhibits GS in vivo (Lamar and Sellinger, 1965), the recovery of activity is most likely dependent on the recruitment of newly generated enzyme, indicating a mechanism counteracting irreversible GS-inhibition.

### Inhibition of GS does not affect glial glutamate transporter quantities

GS-inhibition in slices or prolonged removal of glutamine from neuronal cultures both fail to decrease vesicular glutamate release (Kam and Nicoll, 2007). Persistent release of glutamate but concomitant lack of its recycling via GS might result in enhanced glial glutamate transporter expression, e.g. to prevent accumulation of glutamate in the synaptic cleft. Thus, we measured the relative abundance of GLAST (or EAAT-1) and Glt-1 (or EAAT-2) transporters in cortical preparations. We did not find significant alterations of GLAST (Fig. 2) or Glt-1 (Fig. 3) protein quantities 24h or 72h after MSO treatment, indicating that a stable glutamate release without glial recycling does not result in enhanced expression of glial glutamate transporters.

### Density, but not affinity of benzodiazepine binding sites is affected by GS-inhibition

We quantified regional binding site densities of glutamate, GABA, adenosine and dopamine receptor subtypes using $^3$H-autoradiography. MSO application did not affect binding sites of glutamate (AMPA, kainate, NMDA), adenosine ($A_1$) or dopamine ($D_1$) receptors (Fig. 4) in the investigated regions. In addition, $GABA_A$ as well as $GABA_B$ receptor densities were unchanged after MSO treatment. However, BZ binding was significantly reduced in the piriform and entorhinal cortices as well as in the amygdala 24h and 72h after MSO treatment. Additionally, BZ binding was significantly reduced in the dentate gyrus 24h after MSO treatment. To elucidate whether BZ binding site affinity had changed due to MSO treatment, saturation analysis of whole brain preparations (without cerebellum) were performed (Fig. 5). The maximum binding capacity ($B_{max}$) of membrane preparations was significantly decreased 72h (-23%; $p < 0.01$), but not 24h following application of MSO (Table 1). However, the affinity ($K_D$) of binding sites remained unchanged at both time points. Thus, the regional decrease of binding site densities is not associated with altered binding site affinity. These data suggest that inhibition of the glutamate/glutamine cycle changes GABAergic receptor modulation by reducing BZ binding. However, densities of $GABA_A$ receptors were unchanged, suggesting alterations of the specific receptor subunit composition.

### $GABA_A$ $\alpha_1$ and $\gamma_2$ subunit quantities are differentially altered at distinct time points after GS-inhibition

The vast majority of benzodiazepine-sensitive $GABA_A$ receptors contain both $\alpha_1$ and $\gamma_2$ subunits (Rudolph and Möhler, 2004). Additionally, phosphorylation of serine residues by protein kinase C modulates $\gamma_2$ subunit function (Krishek et al., 1994). To investigate the effects of GS-inhibition on GABAergic receptor composition, western blot analysis were performed for $GABA_A$ subunits $\alpha_1$ (Fig. 6), $\gamma_2$ and phosphorylated $\gamma_2$ (Fig. 7).

Inhibition of GS-activity significantly decreased $\alpha_1$ subunit quantities 24h after treatment (-45%, $p < 0.002$). The same tendency could be observed 72h after MSO application (-28%), although it did not reach significance, indicating a gradual recovery of $\alpha_1$ subunit expression. The protein quantities of the $\gamma_2$ subunit were significantly decreased after 24h (-34%, $p < 0.03$), but increased (+58%, $p < 0.03$) after 72h of GS-inhibition. However, phosphorylated $\gamma_2$ (p-$\gamma_2$) subunit was decreased 24h after MSO treatment (-14%, $p < 0.05$) but did not show significant changes after 72h. Thus, inhibition of glutamate/glutamine cycle differentially affects $GABA_A$ subunit expression.

### Inhibition of GS does not affect $GABA_A$ $\alpha_1$ and $\gamma_2$ subunit mRNA expression

To further investigate whether alterations of GABAergic receptor subunits are based on changes of the respective mRNA transcription levels, we measured the $\alpha_1$ and $\gamma_2$ subunit mRNAs by quantitative in situ hybridization. For both subunits, we did not find significant changes of mRNA expression following MSO treatment (supplementary figure 2), suggesting that the reduction of subunit protein levels described above is not mediated via transcriptional regulation.

**Figure 1.** MSO induces long term inhibition of GS-activity in vivo. Relative GS-activity was measured by γ-glutamyl transferase reaction in cerebral homogenates of rats 24h and 72h after treatment with either a single dose of MSO (100 mg/kg) or vehicle. After 24h MSO reduced GS-activity in treated rats (n = 6) to 30 ± 5 % of controls (n = 6). When measured 72h after treatment (n = 6) GS-activity was reduced to 60 ± 9 % of controls (n = 6).

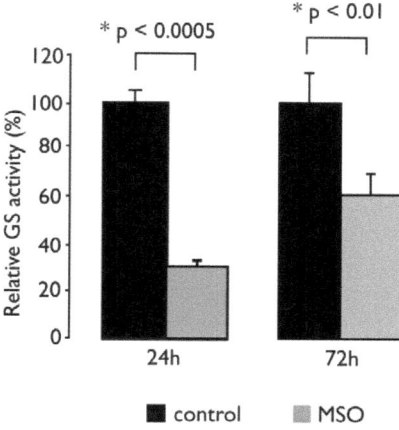

**Figure 2.** Inhibition of the glutamate/glutamine cycle does not change relative GLAST expression **A**, Western Blot of GLAST expression 24h and 72h after treatment with either a single dose of MSO (100 mg/kg) or vehicle. **B**, Bar chart of relative GLAST expression as indicated in A (n = 3 animals per group). GAPDH, glyceraldehyde 3-phosphate dehydrogenase; n.s., not significant.

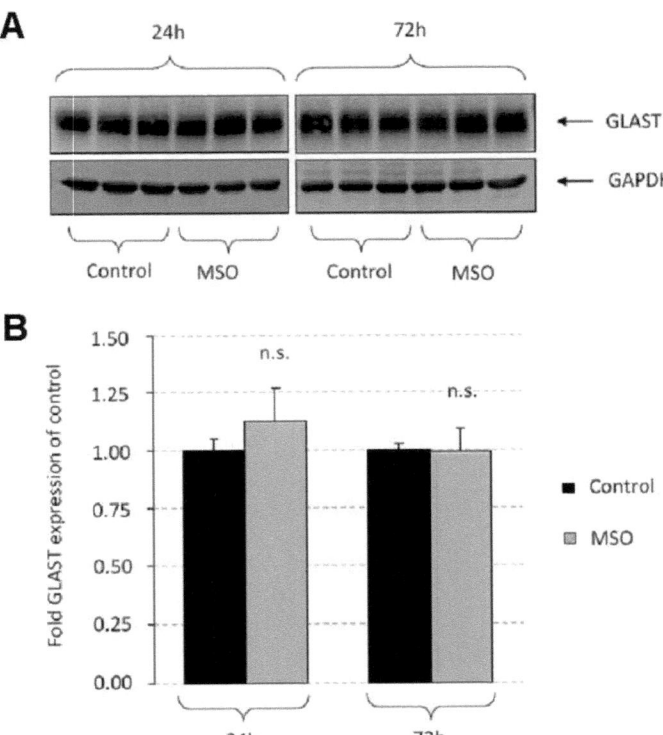

**Figure 3.** Inhibition of the glutamate/glutamine cycle does not change relative Glt-1 expression **A**, Western Blot of Glt-1 expression 24h and 72h after treatment with either a single dose of MSO (100 mg/kg) or vehicle. **B**, Bar chart of relative Glt-1 expression as indicated in A (n = 3 animals per group). GAPDH, glyceraldehyde 3-phosphate dehydrogenase; n.s., not significant.

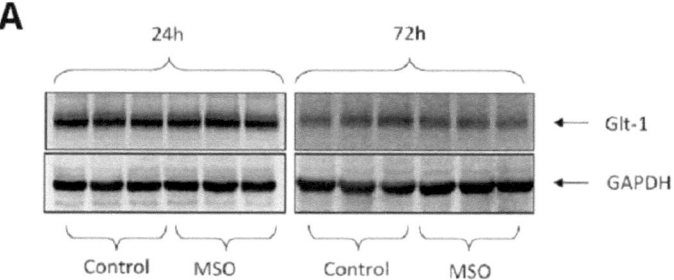

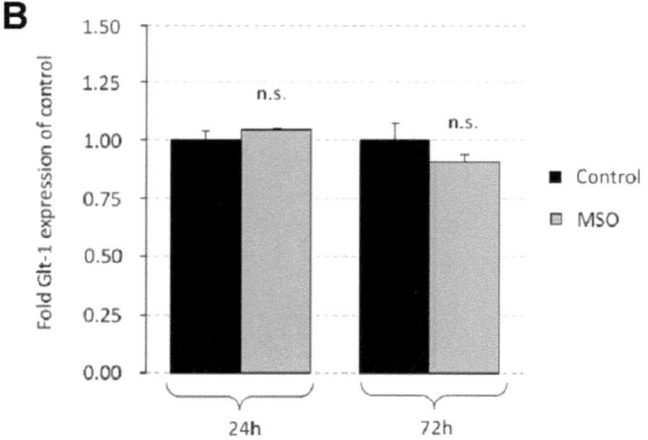

**Figure 4.** Inhibition of glutamate/glutamine cycle decreases BZ binding site densities in different regions of the rat brain. Receptor densities were measured by quantitative $^3$H-autoradiography in rats treated with either a single dose of MSO or vehicle (n = 6 animals per group). A significant decrease of BZ binding site densities was measured in the amygdala and in the piriformal and entorhinal cortices 24h and 72h after treatment. In the dentate gyrus BZ binding was significantly reduced 24h after MSO treatment. Receptor density is given as fmol/mg protein (mean + SEM). Statistically significant changes are marked by asterisks (*p < 0.05; **p < 0.01; ***p < 0.001). RSG, retrosplenial granular cortex; DG, dentate gyrus; Par1, somatosensory cortex; Amg, amygdala; Pir, piriformal cortex; Ent, entorhinal cortex; CP, caudate putamen; SN, Substantia nigra.

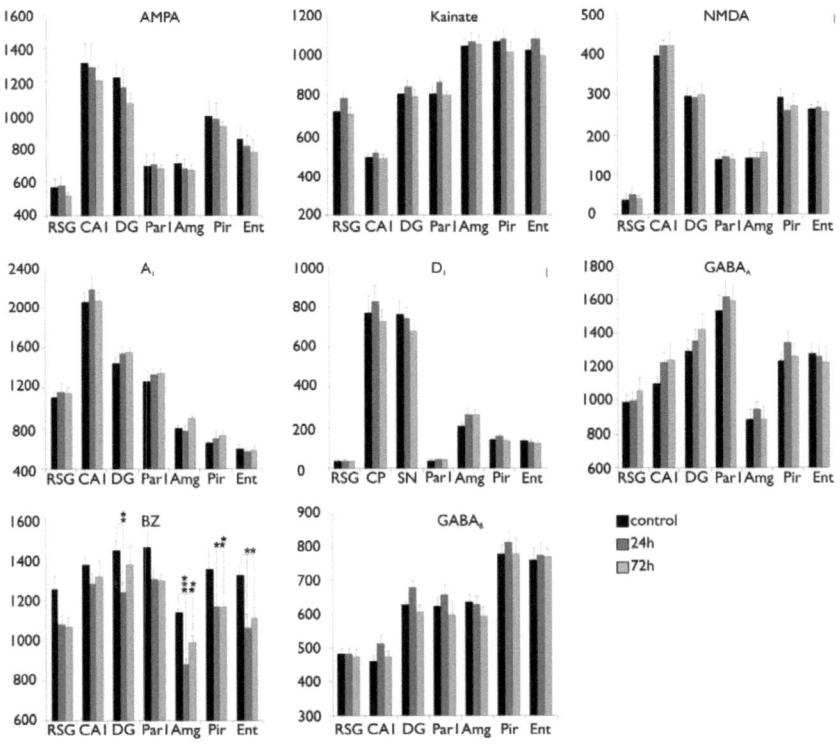

# Inhibition of glutamine synthetase

**Figure 5.** MSO decreases BZ binding site densities, but does not change their affinity 72h after treatment. Representative results of saturation analysis are shown. Left, saturation analysis using non-linear regression. Right, Scatchard plots of the respective data. Control, $B_{max}$ = 2202 fmol/mg protein, $K_D$ = 2.13 nM, r = 0.997. MSO, $B_{max}$ = 1677 fmol/mg protein, $K_D$ = 2.16 nM, r = 0.998. For all saturation analysis unspecific binding was less than 5% (not shown). B, binding; F, ligand concentration.

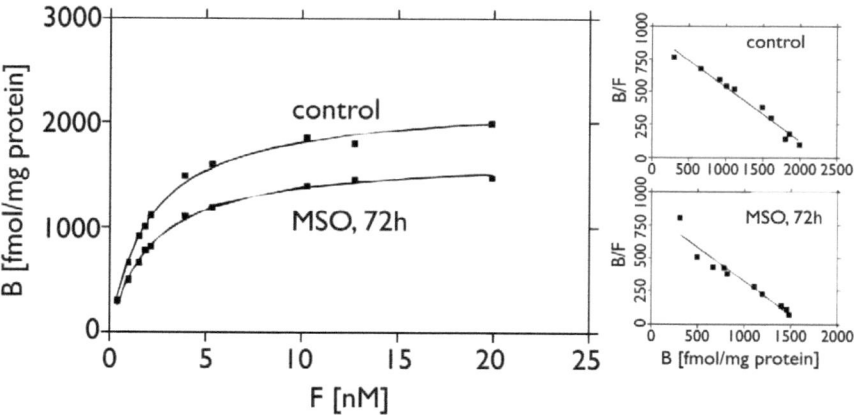

**Table 1.** Mean values of $B_{max}$ and $K_D$ for BZ binding sites in rat brains after treatment with either a single dose of MSO (n = 12) or vehicle (n = 5). Application of MSO significantly reduced $B_{max}$ 72h after treatment (p < 0.01, n = 7), confirming the results measured by receptor autoradiography. Affinity of binding sites remained unchanged by MSO-treatment. $B_{max}$ [fmol/mg protein], $K_D$ [nM].

|  | control | 24h | 72h |
|---|---|---|---|
| $B_{max}$ | 2214 ± 183 | 2169 ± 216 | *1676 ± 78 |
| $K_D$ | 2.18 ± 0.19 | 2.11 ± 0.25 | 2.06 ± 0.09 |

**Figure 6.** Inhibition of glutamate/glutamine cycle decreases relative $GABA_A$ $\alpha_1$ subunit quantities. Western blot analysis of brain homogenates 24h and 72h after MSO treatment. After 24h $\alpha_1$ subunit density is significantly reduced when compared to controls, while the same effect was not found to be significant after 72h (n = 3 per group). GAPDH, glyceraldehyde 3-phosphate dehydrogenase; n.s., not significant.

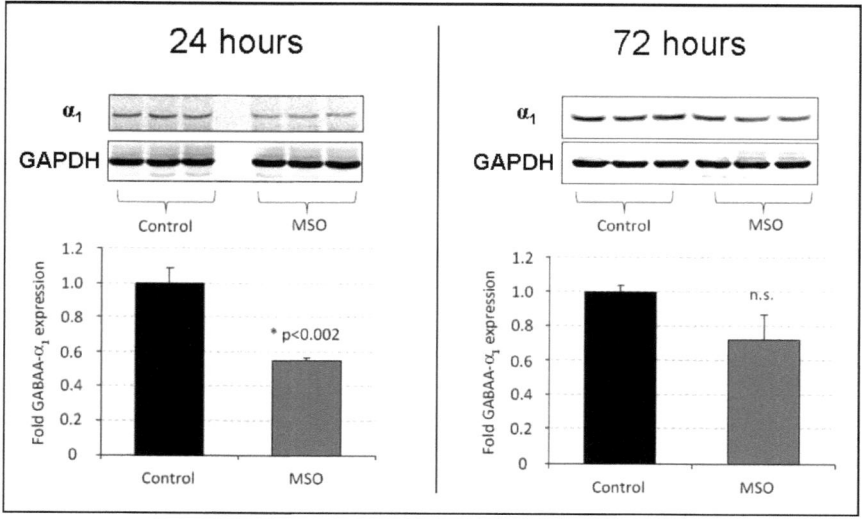

**Figure 7.** Inhibition of glutamate/glutamine cycle differentially affects the relative $GABA_A$ subunit quantities of $\gamma_2$ and its phosphorylated form after 24h and 72h. Western blot analysis of brain homogenates 24h and 72h after MSO treatment. After 24h $\gamma_2$ subunit density is significantly reduced when compared to controls, but increased after 72h. However, serine-phosphorylated $\gamma_2$ subunits were decreased after 24h, but did not show significant changes after 72h (n = 3 per group). GAPDH, glyceraldehyde 3-phosphate dehydrogenase; n.s., not significant.

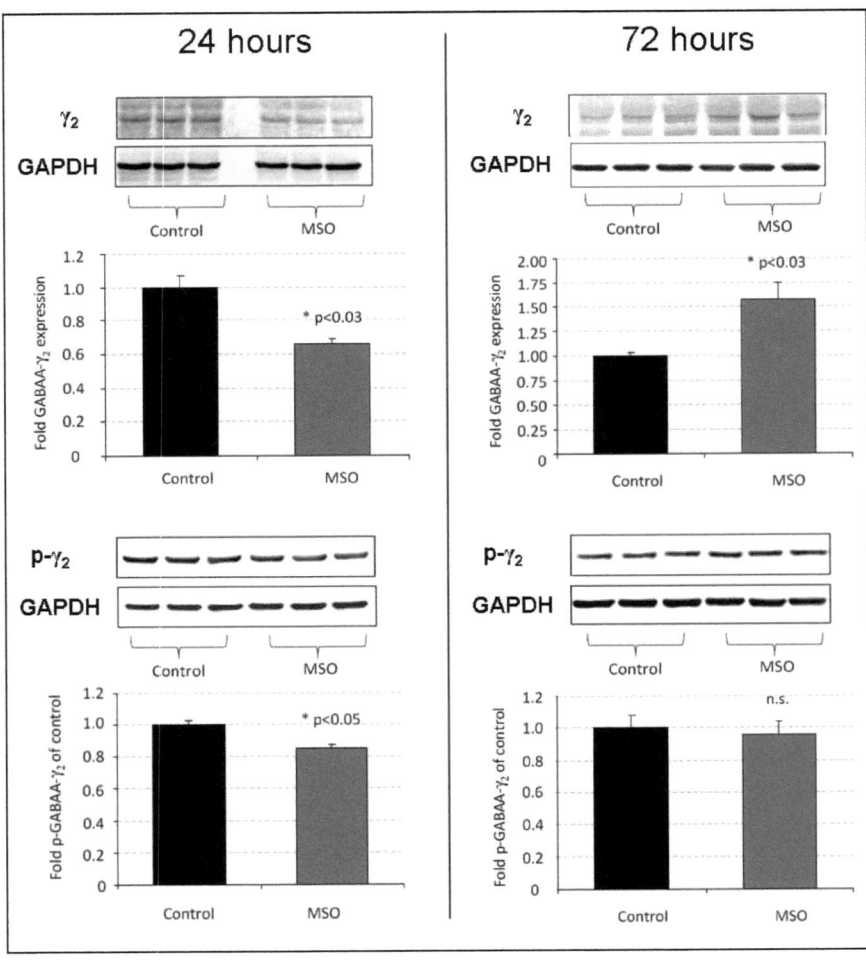

## Discussion

This is the first study to address the effects of MSO-induced in vivo inhibition of astrocytic glutamine synthetase on the densities, affinity and subunit composition of neurotransmitter receptors and the expression of glial glutamate transporters in the rat brain. Inhibition of GS by MSO resulted in decreased BZ binding in distinct brain regions and brain homogenates, but unaltered binding site affinities. During the course of GS-inhibition we found differential changes in the expression of GABA$_A$ receptor subunits $\alpha_1$, $\gamma_2$, and p-$\gamma_2$.

### The glutamate/glutamine cycle regulates GABAergic neurotransmission

In vivo inhibition of the GGC results in decreased GABA content in rat brain homogenates (Stransky, 1969). These results were confirmed by Fonnum and Paulsen (1990) for the neostriatum and they further showed that a substantial proportion of the glutamine pool is linked to GABA metabolism, while Hevor et al. (1996) reported only a slight reduction of GABA levels for the hippocampus. Recently, Liang et al. (2006) demonstrated that the GGC is a major contributor to synaptic GABA release under physiological conditions. Together, these data show that the GGC is an important regulator of GABA homeostasis. However, it remains a matter of debate how GS-inhibition by MSO differentially affects GABA contents in various brain regions.

Binding of endogenous or exogenous BZs to the GABA$_A$ receptor enhances its function. A decrease of synaptic GABA levels could attenuate inhibitory neurotransmission. Thus, one expects an increase in BZ binding to compensate for reduced GABAergic activity or convulsive effects of MSO, as indicated in the pentylenetetrazole model (Cremer et al. 2009a). In contrast, our data demonstrate a decrease of BZ binding and further implies an altered GABAergic modulation due to GS-inhibition. Thus, alterations of GABA$_A$ subunit composition which result in a decrease of BZ binding could occur in an attempt to attenuate the effects of lowered GABA levels and reduced BZ binding. The GABAergic system plays an important role in coping with MSO-induced convulsions as implicated by treating mice with diazepam prior to MSO-administration, which increases seizure latency in a dose dependent manner (Gill and Schatz, 1985). Furthermore, intranigral application of the GABA elevating drug γ-vinyl GABA suppressed MSO-induced seizures (Toussi et al.,

1987). Concerning our present results, both studies underline the functional connection between GGC and GABAergic modulation.

In our model, the first behavioural changes of treated rats appeared approximately eight hours after MSO-application. Our first measurement, however, was performed after 24h, and resulted in a regional reduction of BZ binding. Additionally, a global decrease of BZ binding in brain homogenates was measured 72h after treatment, a time point when animals did not show apparent behavioural alterations from controls. Thus, it seems unlikely that a decrease of BZ binding site densities exclusively accounts for the in vivo effects of MSO. However, it may be assumed that a disturbed GABAergic modulation by endogenous BZs essentially contributes to GGC-associated pathologies.

To conclusively elucidate whether changes of neurotransmitter receptor expression observed in the present study are due to an inhibition of GS itself or secondary to seizure onset additional experiments will be considered. Combined application of MSO together with anticonvulsants to block seizures should offer evidence to investigate this topic. Other ways of GS inhibition (e.g. knock-down of GS expression) would furthermore exclude the possibility of MSO-mediated effects other than GS inhibition.

Regional patterns of $GABA_A$ receptor alterations following GS-inhibition

In MSO treated rats changes of BZ binding site densities and GABAergic subunit composition comprise a regional and temporal dependency. Thus, inhibition of the GGC does not result in a distinct alteration of GABAergic receptors, but is moreover regulated by spatial and temporal factors, the most likely of which is seizure susceptibility (within the epileptic circuitry). Results from autoradiography experiments exhibited decreased densities of BZ binding sites 24h and 72h after MSO treatment. This decrease only proved to be significant in the dentate gyrus 24h after treatment, and after 24h as well as 72h in the amygdala, the piriform and entorhinal cortices. Interestingly, the piriform and entorhinal cortices are key regions within the epileptic circuitry (White, 2002). Since MSO affects all astrocytes, this regional specificity indicates that MSO-induced seizures differentially affect the neuron-glia interaction in seizure prone cerebral regions. Interestingly, this pattern basically corresponds to an enhanced expression of HSP 27 (a label for affected astrocytes) in MSO treated rats and, further, to enhanced GS-nitration following pentylenetetrazole (GABA-antagonist)-induced seizures (Bidmon et al., 2008). Therefore, we propose that the priform/entorhinal cortices, dentate gyrus and hippocampal CA1 region are key regions in which impaired astrocytic glutamate metabolism causes region-specific changes of neurotransmitter receptors in seizure prone neurons and/or glial cells in both animal models. Despite GS-nitration and its reduced activity, BZ binding sites were found to be increased rather than decreased in the respective regions in the PTZ-model (Cremer et al., 2009a). However, this might by due to the fact that MSO and PTZ differentially act on GABAergic neurotransmission. Furthermore, PTZ-treated animals exhibited recurrent seizures and changes of several receptor systems, which most likely influence GABAergic receptors differently (Bidmon et al., 2005; Cremer et al., 2009a).

However, the general assumption of a regional denpendency of MSO effects is supported by a number of studies. Intraventricular but not intrastriatal infusion of MSO leads to behavioural convulsions (Rothstein and Tabakoff, 1982). The protective effect of γ-vinyl GABA against MSO-induced seizures depends on the site of injection (Toussi et al., 1987) and differential changes of amino acid contents have been described for the neostriatum and globus pallidus (Fonnum and Paulsen, 1990) after MSO treatment, while the levels of choline and acetylcholine were differentially and dose dependently altered in the striatum, thalamus, hypothalamus and corpus callosum (Richard and Hevor, 1995).

## Temporal dependency of GABA$_A$ subunit composition following GS-inhibition

During the course of GS-inhibition we found decreased levels of $α_1$ subunits after 24h but not after 72h. Furthermore, we measured a decrease of $γ_2$ and p-$γ_2$ subunits after 24h, but an increase of $γ_2$ without changes of p-$γ_2$ after 72h. Conversely, quantities of functional BZ binding sites in brain homogenates were unchanged after 24h but significantly decreased after 72h. Thus, it seems most likely that subunit expression alterations precede functional binding site alterations at the receptor. As a hypothesis, BZ binding could be increased later than 72h after MSO treatment due to the enhanced expression of $γ_2$ subunits. A follow up study is planned to comprehensively elucidate the temporal patterns of GABA$_A$ subunits and their respective mRNAs following GS-inhibition. Furthermore, we will address the question, whether changes of subunit expression or BZ binding sites can be quantified before behavioural alterations occur or seizures emerge in MSO-treated animals.

$GABA_A$ receptor subunit expression and ligand binding are differentially altered in rodent models of temporal lobe epilepsy (e.g. Volk et al., 2006; Bethmann et al., 2008), and these changes are closely connected to antiepileptic drug resistance (reviewed by Löscher 2009; Schmidt and Löscher 2009). It is tempting to speculate that $GABA_A$ subunit changes induced by GGC disruption might be key events for both epileptogenesis (Eid et al., 2004; 2008) as well as antiepileptic drug resistance. However, further effort is neccesary to verify GS as a potential target for clinical intervention in temporal lobe epilepsy.

Glutamate transport and neurotransmitter receptors

The astrocytic glutamate/glutamine cycle is generally believed to restore the neuronal pool of vesicular glutamate (Kandel et al., 2000). However, Kam and Nicoll (2007) demonstrated that the synaptic release of glutamate remains stable after GS-inhibition in slices or after prolonged removal of glutamine from pure neuronal cultures. A persistent release of glutamate but concomitant lack of its recycling could lead to enhanced concentrations of glutamate in the synaptic cleft, ultimately inducing cell death (e.g. via excitotoxicity). However, we did not find evidence of enhanced cell death in silver-stained histological sections in our model. This observation is supported by experiments on the effect of MSO (350 mg/kg) on blood brain barrier permeability, where ultrastructural investigation demonstrated that MSO, even in high concentrations, does not affect glial integrity (Nitsch et al., 1986) and pathological changes in gial cells are induced only after repeated application of MSO. Additionally, pronounced cellular degeneration in our model would lead to a decrease of receptor densities for several neurotransmitter systems. Beside alterations of the $GABA_A$/BZ system we did not find such changes of receptor densities (cf. fig. 4). Due to a lack of enhanced cell death, we assumed that an increase of synaptic glutamate levels could be prevented by enhanced glutamate reuptake. The astrocytic membrane is responsible for at least 80% of glutamate clearance and the majority of synaptic inactivation in the brain (Bergles and Jahr, 1997; Danbolt, 2001). Astrocytic glutamate uptake is significantly enhanced for up to 7 days after intraventricular injection of MSO (Rothstein and Tabakoff, 1984). Therefore, we hypothesized that an enhanced glial glutamate transport might lead to an increased transporter expression. GLAST and Glt-1 are the predominant glutamate transporters in the brain and both are mainly expressed by astrocytes (Gadea and López-Colomé, 2001). However, we did not find changes of either GLAST or Glt-1 expression. Thus, available glutamate transporters sufficiently remove synaptic glutamate, or an enhanced transport is not mediated by increased transporter expression, but rather by changes in removal rate or affinity (Rothstein and Tabakoff, 1984).

Intraventricular as well as intrastriatal injection of MSO results in a transient increase in striatal dopamine release followed by an inhibition of its release for up to three days (Rothstein and Tabakoff, 1982). Apomorphine, a $D_1/D_2$ receptor agonist, produces a concentration related rise in glutamate concentration in cerebral perfusates, which is abolished by intrastriatal MSO injection. Together, these data imply a close correlation between the GGC and dopaminergic regulation of glutamate release.

In our model, GS-inhibition did not change the densities of the glutamatergic AMPA-, kainate- and NMDA-receptors nor of dopaminergic $D_1$ receptors. Additionally, adenosine $A_1$ receptors (important negative regulators of vesicular glutamate release, e.g. Fredholm et al., 2005) were unaffected by GS-inhibiton. Thus, we suggest that neither changes of $D_1$, nor the glutamatergic subtypes AMPA, kainate or NMDA significantly contribute to an altered neurotransmission following GS-inhibition. Furthermore, it is implicated that regulation of vesicular glutamate release by adenosine receptors is independent of the GGC.

The astrocytic enzyme glutamine syntethase is a key regulator of glutamate and GABA metabolism. Here, we have shown that inhibition of GS by MSO leads to changes of GABAergic but not glutamatergic neurotransmitter receptors or modulators. In conclusion, our results suggest a regulatory circuit between the GGC and GABAergic synaptic transmission, which is mediated by changes of $GABA_A$ subunit composition and altered modulation by BZs.

**Acknowledgments**

This study was partially supported by a grant of the Helmholtz Alliance in "Mental Health in an Ageing Society". We wish to thank S. Buller, L. Igdalova and S. Wilms for excellent technical assistance.

**Conflict of interest:** We confirm that we have read the Journal's position on issues involved in ethical publication and affirm that this report is consistent with those guidelines. None of the authors has any conflict of interest to disclose.

**Supplementary figure 1.** Schematic definition of ROIs at (A) Bregma -3.3 mm and (B) Bregma -5.3 mm according to the cytoarchitectonic atlas of Zilles (1985). AMG, amygdala; DG, dentate gyrus; Ent, entorhinal cortex; Par1, somatosensory cortex; Pir, piriform cortex, PRh, perirhinal cortex, Put, putamen; RSG, retrosplenial granular cortex; SN, substantia nigra.

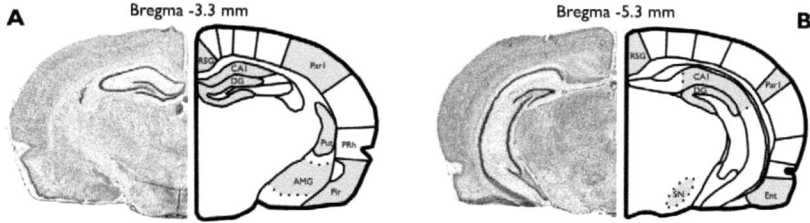

**Supplementary figure 2.** MSO treatment does not affect the mRNA levels of $\alpha_1$ or $\gamma_2$ $GABA_A$ subunits. Results from quantitative in situ hybridization in ROIs of rats treated with either a single dose of MSO or vehicle (n = 4-6). mRNA density is given as Bq/mg tissue + SEM.

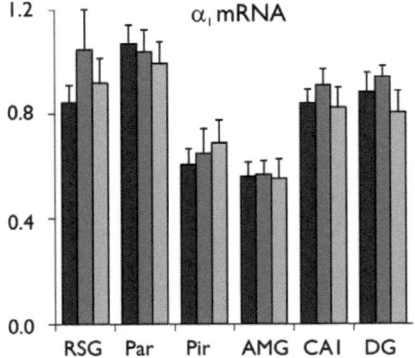
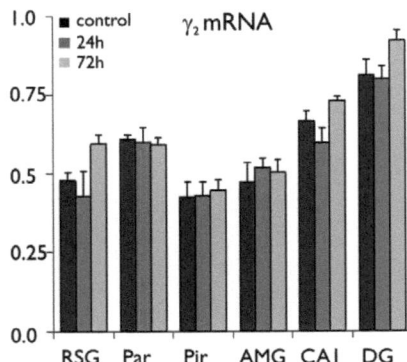

**References**

Bak LK, Schousboe A, Waagepetersen HS (2006) The glutamate/GABA-glutamine cycle: aspects of transport, neurotransmitter homeostasis and ammonia transfer. J Neurochem 98:641-653.

Basile A (1997) Saturation assays of radioligand binding to receptors and their allosteric modulatory sites. In: Current Protocols in Neuroscience (Crawley, Gerfen, Rogawsky, Sibley Skolnick, Wray, eds), chapter 7.6, New York: John Wiley and Sons, Inc.

Bethmann K, Fritschy JM, Brandt C, Löscher W (2008) Antiepileptic drug resistant rats differ from drug responsive rats in GABA A receptor subunit expression in a model of temporal lobe epilepsy. Neurobiol Dis. 31:169-87.

Bergles DE, Jahr CE (1997) Synaptic activation of glutamate transporters in hippocampal astrocytes. Neuron 19:1297-1308.

Bidmon HJ, Görg B, Palomero-Gallagher N, Schleicher A, Häussinger D, Speckmann EJ, Zilles K.(2008) Glutamine synthetase becomes nitrated and its activity is reduced during repetitive seizure activity in the pentylentetrazole model of epilepsy. Epilepsia 49:1733-1748.

Bidmon HJ, Görg B, Palomero-Gallagher N, Schliess F, Gorji A, Speckmann EJ, Zilles K. (2005) Bilateral, vascular and perivascular glial upregulation of heat shock protein-27 after repeated epileptic seizures. J Chem Neuroanat 30:1-16.

Chaudhry FA, Reimer RJ, Edwards RH (2002) The glutamine commute: take the N line and transfer to the A. J Cell Biol 157:349-355.

Cremer CM, Palomero-Gallagher N, Bidmon HJ, Schleicher A, Speckmann EJ, Zilles K. (2009a) Pentylenetetrazole-induced seizures affect binding site densities for GABA, glutamate and adenosine receptors in the rat brain. Neuroscience 163:490-9.

Cremer CM, Cremer M, Lopez Escobar J, Speckmann EJ, Zilles K. (2009b) Fast, quantitative in situ hybridization of rare mRNAs using (14)C-standards and phosphorus imaging. J Neurosci Methods 185:56-61.

Danbolt NC (2001) Glutamate uptake. Prog Neurobiol 65:1-105

Eid T, Ghosh A, Wang Y, Beckström H, Zaveri HP, Lee TS, Lai JC, Malthankar-Phatak GH, de Lanerolle NC (2008) Recurrent seizures and brain pathology after inhibition of glutamine synthetase in the hippocampus in rats. Brain 131:2061-2070.

Eid T, Thomas MJ, Spencer DD, Rundén-Pran E, Lai JC, Malthankar GV, Kim JH, Danbolt NC, Ottersen OP, de Lanerolle NC (2004) Loss of glutamine synthetase in the human epileptogenic hippocampus: possible mechanism for raised extracellular glutamate in mesial temporal lobe epilepsy. Lancet 363:28-37.

Fonnum F, Paulsen RE (1990) Comparison of transmitter amino acid levels in rat globus pallidus and neostriatum during hypoglycemia or after treatment with methionine sulfoximine or gamma-vinyl gamma-aminobutyric acid. J Neurochem 54:1253-1257.

Fredholm BB, Chen JF, Masino SA, Vaugeois JM (2005) Actions of adenosine at its receptors in the CNS: insights from knockouts and drugs. Annu Rev Pharmacol Toxicol 45:385-412.

Gadea A, López-Colomé AM (2001) Glial transporters for glutamate, glycine and GABA I. Glutamate transporters. J Neurosci Res 63:453-60.

Gill MW, Schatz RA (1985) The effect of diazepam on brain levels of S-adenosyl-L-methionine and S-adenosyl-L-homocysteine: possible correlation with protection from methionine sulfoximine seizures. Res Commun Chem Pathol Pharmacol 50:349-363.

Hevor TK (1996) Methionine sulfoximine has no major effect on gamma-aminobutyric acid concentration in the rat brain. Biog Amines 12:445-461.

Kam K, Nicoll R (2007) Excitatory synaptic transmission persists independently of the glutamate-glutamine cycle. J Neurosci 27:9192-9200.

Kandel E, Schwartz J, Jessell T (2000) Principles of neural science, Ed 4. New York: McGraw-Hill

Krishek BJ, Xie X, Blackstone C, Huganir RL, Moss SJ, Smart TG (1994) Regulation of GABAA receptor function by protein kinase C phosphorylation. Neuron 12:1081-1095.

Kvamme E, Roberg B, Torgner IA (2000) Phosphate-activated glutaminase and mitochondrial glutamine transport in the brain. Neurochem Res 25:1407-1419.

Lamar C Jr, Sellinger OZ (1965) The inhibition in vivo of cerebral glutamine synthetase and lutamine transferase by the convulsant methionine sulfoximine. Biochem Pharmacol 14:489-506

Liang SL, Carlson GC, Coulter DA (2006) Dynamic regulation of synaptic GABA release by the glutamate-glutamine cycle in hippocampal area CA1. J Neurosci. 26:8537-8548.

Löscher W. (2009) Molecular mechanisms of drug resistance in status epilepticus. Epilepsia 50 Suppl 12:19-21.

Martin DL, Tobin AJ (2000) Mechanisms controlling GABA synthesis and degradation in the brain. In: GABA in the nervous system (Martin DL, Olsen RW, eds), pp 25-41. Philadelphia: Lippincott Williams & Wilkins.

Martinez-Hernandez A, Bell KP, Norenberg MD (1977) Glutamine synthetase: glial localization in brain. Science 195:1356-1358.

Nitsch C, Goping G, Klatzo I. (1986) Pathophysiological aspects of blood-brain barrier permeability in epileptic seizures. Adv Exp Med Biol. 203:175-89.

Palomero-Gallagher N, Bidmon H-J, Cremer M, Schleicher A, Kircheis G, Reifenberger G, Kostopoulos G, Häussinger D, Zilles K (2009) Neurotransmitter receptor imbalances in motor cortex and basal ganglia in hepatic encephalopathy. Cellular Physiology and Biochemistry 24:291-306.

Rao VL, Murthy CR (1988) Age-dependent variation in the sensitivity of rat brain glutamine synthetase to L-methionine-DL-sulfoximine. Int J Dev Neurosci 6:425-430.

Richard O, Hevor T (1995) Methionine sulfoximine increases acetylcholine level in the rat brain: no relation with epileptogenesis. Neuroreport 6:2027-2032.

Rodrigo R, Felipo V. (2007) Control of brain glutamine synthesis by NMDA receptors. Front Biosci 12:883-890.

Ronzio RA, Rowe WB, Meister A (1969) Studies on the mechanism of inhibition of glutamine synthetase by methionine sulfoximine. Biochemistry 8:1066-1075.

Rothstein JD, Tabakoff B (1982) Effects of the convulsant methionine sulfoximine on striatal dopamine metabolism. J Neurochem 39:452-457.

Rothstein JD, Tabakoff B (1984) Alteration of striatal glutamate release after glutamine synthetase inhibition. J Neurochem 43:1438-46.

Rothstein JD, Tabakoff B (1985) Glial and neuronal glutamate transport following glutamine synthetase inhibition. Biochem Pharmacol 34:73-79.

Rudolph U, Möhler H (2004) Analysis of GABAA receptor function and dissection of the pharmacology of benzodiazepines and general anesthetics through mouse genetics. Annu Rev Pharmacol Toxicol 44:475-498.

Schmidt D, Löscher W (2009) New developments in antiepileptic drug resistance: an integrative view. Epilepsy Curr. 9:47-52.

Stransky Z (1969) Time course of rat brain GABA levels following methionine sulphoximine treatment. Nature 224:612-613.

Toussi HR, Schatz RA, Waszczak BL (1987) Suppression of methionine sulfoximine seizures by intranigral gamma-vinyl GABA injection. Eur J Pharmacol 137:261- 264.

Volk HA, Arabadzisz D, Fritschy JM, Brandt C, Bethmann K, Löscher W (2006) Antiepileptic drug-resistant rats differ from drug-responsive rats in hippocampal neurodegeneration and GABA(A) receptor ligand binding in a model of temporal lobe epilepsy. Neurobiol Dis. 21:633-46.

Webb JT, Brown GW (1976) Some properties and occurrence of glutamine synthetase in fish. Comp Biochem Physiol B 54:171-175.

White HS (2002) Animal models of epileptogenesis. Neurology 59: 7-14.

Yamamoto T, Iwasaki Y, Sato Y, Yamamoto H, Konno H (1989) Astrocytic pathology of methionine sulfoximine-induced encephalopathy. Acta Neuropathol 77:357-368.

Zilles K (1985) The cortex of the rat – a stereotaxic atlas, Berlin Heidelberg: Springer

Zilles K, Schleicher A (1991) Quantitative receptor autoradiography and image analysis. Bull Assoc Anat (Nancy ) 75:117-121.

Zilles K, Schleicher A (1995) Correlative imaging of transmitter receptor distributions in human cortex. In: Autoradiography and Correlative Imaging (Stumpf WE, Solomon HF, eds), pp 277-307. San Diego: Academic Press.

Zilles K, Schleicher A, Palomero-Gallagher N, Amunts K (2002) Quantitative analysis of cyto- and receptor architecture of the human brain. In: Brain Mapping. The Methods (Toga AW, Mazziotta JC, eds), pp 573-602. Amsterdam: Elsevier.

Zilles K, Palomero-Gallagher N, Schleicher A (2004) Transmitter receptors and functional anatomy of the cerebral cortex. J Anat 205:417-432.

**Laminar distribution of neurotransmitter receptors in the *reeler* mouse cerebral cortex**

Christian M. Cremer, Joachim H.R. Lübke, Nicola Palomero-Gallagher, and Karl Zilles

In *reeler* mice a mutation of the extracellular matrix protein *reelin* leads to developmental deficits of neuronal migration causing cerebellar hypoplasia, disturbed laminar pattern of the hippocampus and an inversion of neocortical layers. In the adult brain, *reelin* is suggested to regulate synaptic plasticity by modulating neurotransmitter receptor function. Little is known, however, about the density and distribution pattern of neurotransmitter receptors in *reeler* mice. Thus, we comprehensively examined different neurotransmitter receptors in *reeler* and wild type brains.

Our data demonstrate differential changes in the laminar distribution, maximum binding capacity ($B_{max}$) and regional density of several neurotransmitter receptors in the *reeler* brain. Thus, these results support the hypothesis for a role of reelin in neurotransmitter receptor regulation.

Contributions on experimental design, realization and publication:
The autoradiographic experiments, saturation analysis and histological stainings as well as the analysis of the respective results were performed by C. Cremer. All technical assistance is acknowledged in the manuscript. The manuscript including all figures and tables was prepared by C. Cremer and subsequently reviewed, amended and approved by all co-authors.

Approximated total share of contribution in per cent: 75%

Current state of publication: in review, The Journal of Neuroscience

Publications

Senior Editor: Dr. Chris J. McBain
Reviewing Editor: Dr. Mark Farrant

**Laminar distribution of neurotransmitter receptors in the *reeler* mouse cerebral cortex**

Christian M. Cremer[1,2*], Joachim H.R. Lübke[1,3,4*], Nicola Palomero-Gallagher[1], and Karl Zilles[1,2,3]

[1]Institute of Neuroscience and Medicine (INM-2), Research Center Jülich, D-52425 Jülich, Germany; [2]C. & O. Vogt Institute for Brain Research, Heinrich-Heine-University, D-40225 Düsseldorf, Germany; [3]JARA Translational Brain Medicine, Germany; [4]Department of Psychiatry and Psychotherapy, Medical Faculty, RWTH Aachen, Germany

* The first two authors have equally contributed to the work

**Abbreviated title:** Neurotransmitter receptors in *reeler* mice

**Address correspondence to:**   Prof. Joachim Lübke
Institute for Neuroscience and Medicine (INM-2)
Research Centre Jülich
D-52425 Jülich
j.luebke@fz-juelich.de

**Number of figures:**   10
**Number of tables:**   3
**Supplemental material:**   1

**Abstract:**   252 words
**Introduction:**   476 words
**Discussion:**   1489 words

**Keywords:**

*reeler* mouse, neurotransmitter receptors, mapping, quantitative receptor autoradiography, *reelin*, synaptic transmission and plasticity

**Acknowledgements:**

The excellent technical assistance by S. Buller, M. Cremer and S. Wilms is very much appreciated. We further thank Dr. Axel Schleicher for helpful comments on the data. This work was supported by the Initiative and Networking Fund of the Helmholtz Association (Helmholtz Alliance on Mental Health in an Ageing Society).

**Abstract**

Mapping of multiple neurotransmitter receptors provides functionally relevant information about the chemoarchitectonic organization of the brain, since neurotransmitter receptors are key elements of synaptic transmission and plasticity. In *reeler* mice a mutation of the extracellular matrix protein *reelin* leads to developmental deficits of neuronal migration causing cerebellar hypoplasia, disturbed laminar pattern of the hippocampus and an inversion of neocortical layers. In the adult brain, *reelin* is suggested to regulate synaptic plasticity by modulating neurotransmitter receptor function. Little is known, however, about the density and distribution pattern of neurotransmitter receptors in *reeler* mice.

Using quantitative *in vitro* receptor autoradiography we examined different neurotransmitter receptors in *reeler* and wild type brains. In particular, the binding site densities and laminar distribution of various glutamate, GABA, muscarinic and nicotinic acetylcholine, serotonin, dopamine and adenosine receptors were analyzed.

Our data demonstrate differential changes in the laminar distribution, maximum binding capacity ($B_{max}$) and regional density of several neurotransmitter receptors in the *reeler* brain. In the neocortex some receptor subtypes demonstrated an obvious laminar inversion (i.e. AMPA, kainate, NMDA, $GABA_B$, $5-HT_1$, $M_1$, $M_3$, nAch), while other subtypes ($A_1$, $GABA_A$, BZ, $5-HT_2$, $M_2$, $\alpha_1$, $\alpha_2$) were less strikingly affected. A significant decrease of whole brain $B_{max}$ was found for adenosine $A_1$ and $GABA_A$ receptors. In the forebrain several binding site densities were differentially altered (i.e. kainate, $A_1$, benzodiazepine, $5-HT_1$, muscarinic $M_2$, adrenergic $\alpha_1$ and $\alpha_2$). Taken together, our results support the hypothesis for a role of reelin in neurotransmitter receptor regulation. Implications for synaptic plasticity and neurodegenerative disease are discussed.

## Introduction

Neurotransmitter receptors are major determinants of synaptic transmission and plasticity in the brain. They are inhomogeneously distributed throughout the cortex, with both regional and laminar characteristics in their location. Local changes in the density and distribution of receptors coincide with cytoarchitectonically defined areas (for review see Zilles and Amunts, 2009).

The cytoarchitecture of the so called reeler mouse is severely disturbed. During development in these mice, a lack of the secreted glycoprotein reelin causes severe cerebellar hypoplasia (Hamburgh, 1960, 1963), mislamination of the hippocampus (Stanfield and Cowan, 1979; Zhao et al., 2004) and an inversion of cortical layers (Caviness, 1982; reviewed by Lambert de Rouvroit and Goffinet, 1998; D'Arcangelo, 2005). Although the *reeler* mouse mutant is a well established model to study mechanisms of cortico- and synaptogenesis, little is known about the density and distribution pattern of neurotransmitter receptors in these mice.

In contrast to the defined role of *reelin* during development relatively little is known about its function in the adult brain. *Reelin*-signaling influences granule cell dispersion in the dentate gyrus during experimental epilepsy, implying a role in the maintenance of cortical lamination (Heinrich et al., 2006; Müller et al., 2009). Furthermore, an increasing body of evidence suggests a function of *reelin* in synaptic transmission and plasticity. Superfusion of *reelin* in hippocampal slices enhances long-term potentiation mediated by the *reelin* receptors VLDLR and ApoER2 (Weeber et al., 2002). It has also been shown that *reelin* regulates NMDA receptor-mediated neuronal activity by tyrosine-phosphorylation (Chen et al., 2005). Other experiments revealed an enhancement of NMDA- and AMPA receptor-mediated activity following treatment with recombinant *reelin* (Qiu et al., 2006). Furthermore, it has been shown that synaptic plasticity is severely altered after electrical stimulation of the cortico-striatal pathway in *reeler* mice, probably depending on a reduced GABAergic tone (Marrone et al., 2006). Recently, Niu et al. (2008) demonstrated that dendritic spine development is promoted by the *reelin* pathway in the postnatal hippocampus.

These studies support the hypothesis that *reelin* deficiency may be associated with alterations of neurotransmitter receptor densities in the *reeler* mutant mouse. Since we have previously shown that changes of a single receptor type are frequently accompanied by receptor alterations of other neurotransmitter systems (e.g. Zilles et al., 1999; Palomero-Gallagher et al., 2009; Cremer et al., 2009; 2010), we conducted a multi-receptor study to investigate the influence of the *reelin* mutation on various neurotransmitter receptors in the adult brain by means of *in vitro* receptor autoradiography. Furthermore, saturation analysis were performed for each receptor type to determine receptor binding affinity (dissociation constant, $K_D$) and maximum binding capacity ($B_{max}$).

Our data demonstrate significant and differential changes in the laminar distribution pattern, $B_{max}$ and regional density of several neurotransmitter receptors in all investigated regions of the *reeler* brain. The affinity of the different receptor types is not changed in the *reeler* mutant, suggesting that receptor subunit composition remains unaltered.

## Material and Methods

**Animals.** *Reeler*[rl] mutant mice (breeding pairs were kindly provided by Prof. André Goffinet, University of Louvain, Brussels, Belgium) and normal CD-1 mice that served as controls (Charles River, Germany) were kept and bred under standard conditions (light/dark cycle 12:12 h, temperature of 20 ± 2 °C, air humidity 55–60 %). For the present study 12-16 week old male mice were used. All experimental procedures were conducted according to the German Animal Welfare Act and guidelines of the Research Centre Jülich after approval by the governmental agency.

**Autoradiography.** Animals were sacrificed by decapitation between 8:00 and 9:00 a.m. The brains were removed from the skull, immediately frozen in isopentane (-50°C) and stored at -80°C until use. All brains were serially sectioned (10 μm thickness) in the coronal plane with a Cryostat (Leica Systems, Wetzlar, Germany). Sections were thaw-mounted on silanised glass slides. The glutamate receptors AMPA, NMDA and kainate, the GABAergic receptors $GABA_A$, $GABA_B$ and $GABA_A$ associated benzodiazepine (BZ) binding sites, the muscarinic acetylcholine receptors $M_1$, $M_2$ and $M_3$, the nicotinic acetylcholine receptor (nAch), the serotonergic receptors $5-HT_1$ and $5-HT_2$, the dopaminergic receptor $D_1$, the adrenergic $\alpha_1$ and $\alpha_2$ receptors, and the adenosine $A_1$ receptor were labeled with the respective tritiated ligands. The detailed procedure has been described elsewhere (Zilles et al., 2002, 2004). Binding conditions, ligands and displacer are summarized in Table 1.

Briefly, sections were washed in buffer, transferred to solution containing the specific [$^3$H]-ligand and finally rinsed to remove unbound ligand. To monitor and quantify non-specific binding, several sections were incubated with the radiolabeled [$^3$H]-ligand in the presence of a specific, non-radioactive competitor. All brains were processed simultaneously in the same buffers and radioactive solutions. Sections were subsequently air-dried and exposed to ß-sensitive films (BioMax MR Film, Kodak Europe) for 10 to 15 weeks in the presence of [$^3$H]-microscale standards (Amersham Biosciences Europe) with known concentrations of radioactivity. All ligands were purchased from Perkin Elmer (Germany) except for [$^3$H]-CGP 54626 (BioTrend, Germany) and [$^3$H]-CPFPX (Institute for Neuroscience and Medicine – Nuclear Chemistry (INM-4), Research Centre Jülich). Competitors CGP 55845 and SYM 2081 were purchased from Tocris Bioscience (Great Britain), quisqualate, MK 801, GABA, Carbachol and Phentolaminmesylate from Biotrend (Germany). All other competitors were obtained from Sigma Aldrich (Germany).

**Image analysis.** Densitometric measurements of autoradiographic films were performed as described previously (Zilles et al., 2002). In summary, autoradiographic films were developed (Hyperprocessor, Amersham Biosciences Europe), digitized (8-bit coding, 1300 x 1030 pixels) using a CCD-camera (Progres C14, Jenoptik, Germany) equipped with a 55 mm objective (Zeiss, Germany) connected to a KS400®-Software system (Zeiss, Germany). Microscales were used to compute a transformation curve representing the relationship between grey values in the autoradiographs and concentrations of radioactivity in the tissue. Regions of interest (ROIs) were traced according to the adult mouse brain atlas by Paxinos and Franklin (2001). Receptor densities were calculated for the somatosensory cortex (Par 1), motor cortex (Mot), putamen (Put) and the subregions CA1, CA3, and dentate gyrus (DG) of the hippocampus in 3-5 randomized sections of 5-8 brains per group (Fig. 1). ROIs were selected to compare principally different cortical organizations, i.e. a granular versus an agranular isocortex (Par1 vs. Mot), isocortex versus allocortex (Par1 and Mot vs. hippocampal subregions CA1, CA3, and DG), and cortical versus non-layered subcortical structures (Par1, Mot, CA1, CA3 and DG vs. Put).

**Receptor density profiles.** Receptor density profiles from the pial surface to white matter border were extracted in the somatosensory cortex. Briefly, the inner and outer cortical surfaces were manually traced on linearised autoradiographs (using MatLab®, The MathWorks, MA, USA), and traverses vertically oriented to the cortical layers were calculated. Receptor density profiles were extracted along the course of eleven neighboring equidistant traverses, and a normalized mean profile was calculated from 3-5 sections for each receptor and mouse. Together with the color coded autoradiographs of whole sections, profiles were used to identify a potentially inverted receptor distribution in *reeler* mutant mice (Figures 3-10).

**Brain homogenates.** To exclude concomitant changes of the receptor binding site properties by the mutation, saturation analyses were performed for each receptor subtype on purified membranes of whole brain homogenates to determine receptor binding affinity (dissociation constant, $K_D$) and maximum binding capacity ($B_{max}$).

Therefore, animals were sacrificed by decapitation and the brains were immediately removed from the skull, transferred into 10 ml of ice cold tris-citrate buffer (50 mM, pH 7.4) and homogenized for 1 minute at 15.000 rpm (Miccra D-13, Art-Labortechnik, Germany). The resulting homogenates were washed four times by addition of 10 ml tris-citrate buffer and subsequent centrifugation (4°C, 14.800 g, 60 min). All homogenates were stored in aliquots at -80°C.

**Saturation analysis.** Experiments were carried out as recently described (Palomero-Gallagher et al., 2009; Cremer et al., 2010) with some modifications. Binding conditions are summarized in Table 1.

*Slice assay.* Since extremely short washing steps (in the range of seconds) cannot be performed reliably enough, the $K_D$ values of AMPA, kainate, NMDA, $GABA_A$, 5-$HT_1$, $M_1$ and $D_1$ binding sites were determined on cryosectioned pellets. Therefore, homogenates as described above (n = 6 per group) were thawed, centrifuged (4°C, 14.800 g, 60 min) and the resulting pellet was cut with a cryostat into 10 µm thick sections and mounted on silanised glass slides. Homogenate sections were incubated following the same protocols as those used for autoradiographic experiments on the intact brain sections but with 9-12 increasing concentrations of the respective ligands (Table 1). Sections were exposed to phosphor imaging plates (BAS-TR 2025; Raytest-Fuji) for 48-72 hours together with radioactive calibration standards (Microscales, Amersham Biosciences Europe), and processed for densitometry (BAS-5000 reader, Raytest-Fuji). Densitometric measurement was performed using standard imaging software (AIDA 2.31; Raytest).

*Filtration assay.* For use in filtration assays homogenates (n = 6 per group) were thawed, centrifuged (4°C, 14.800 g, 60 min) and the resulting pellet diluted in incubation buffer (Table 1). Protein concentration of homogenates was photometrically determined prior to the experiments (0.3-0.6 mg/ml). Final assay volume was 300 µl. In glass reaction tubes 50 µl aliquots of radiolabeled ligands (9-12 increasing concentrations) were added to 100 µl of brain homogenate and 150 µl of incubation buffer (Table 1). Triplicated measurements were performed for each ligand concentration by scintillation spectrometry using a Beta-Counter (Tri-Carb 2100TR, Packard BioScience). Non-specific binding was monitored by excess addition of specific competitors. Final concentration of protein ranged between 0.1-0.2 mg/ml. In order to quantify filter binding, brain homogenates in the assay were replaced by 100 µl of incubation buffer. Binding was assessed by rapid filtration (< 5 s) over GF/B glass fiber filters pre-soaked for 1 min in 0.03 % polyethylenimine, followed by 3 washes (< 10 s) with 5 ml buffer using a cell harvester (M-48, Brandel, MD, USA). GF/B filters were transferred into scintillation vials containing 10 ml scintillation fluid (Ready Safe™, Beckman Coulter, Germany), left on a rocking plate at room temperature overnight and counted by scintillation spectrometry. For each experiment all homogenates were processed simultaneously using the same buffers, ligands and concentrations of radioactivity.

**Assay analysis.** Specific binding was determined as the difference between total binding and non-specific binding. $K_D$ and $B_{max}$ were determined from the experiments by computerized non-linear regression analysis (PRISM 2.0, GraphPad Software, San Diego, CA, USA). To monitor quality of filtration assay a ratio was calculated determining binding specificity as percent of total binding: (non-specific binding - filter binding) *100 / total binding – filter binding.

**Statistical analysis.** Differences of receptor densities extracted from ROIs were analyzed for statistical significance by one way analysis of variance with repeated measurements (rANOVA; $p < 0.05$). Receptors demonstrating significant differences were further analyzed by two tailed Student's t-test, to discriminate between ROIs. Data from saturation analysis were analyzed using a one way analysis of variance (ANOVA; $p < 0.01$).

## Fig. 1: Regions of interest (ROIs) investigated by receptor autoradiography

Left panel: schematic drawing of a brain section according to the cytoarchitectonic atlas of Paxinos and Franklin (2001) at Bregma -1.8mm. Right panel: Silver stained section at the same level. ROIs are highlighted. Abbreviations: Amg amygdala; CA1 CA1 subregion of the hippocampus; CA3 CA3 subregion of the hippocampus; DG dentate gyrus; E entopenduncular cortex; F fimbria fornix; FL/HL forelimb/hindlimb region; Ins insular cortex; Mot motor cortex; Par1 primary somatosensory cortex; Par2 secondary somatosensory cortex; Put Putamen; Rag retroplenial agranular cortex; Rg retrosplenial granular cortex.

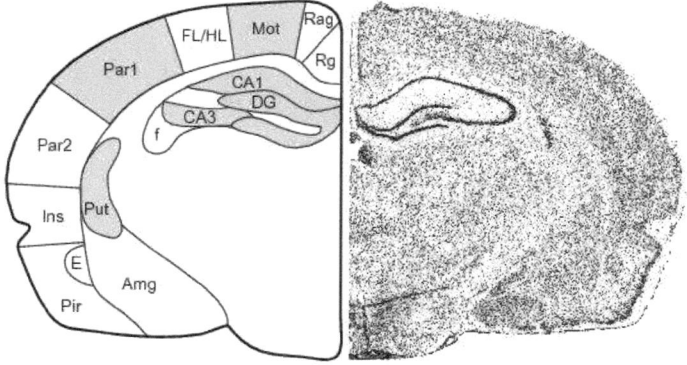

**1: Binding conditions of quantitative receptor autoradiography and saturation analyses.** Autoradiographic experiments were performed according to standard protocols using 10 μm cryosections of both wild type and reeler brains. Saturation analyses were performed using purified membrane preparations of whole brain homogenates that were either used in filtration assay or pelletized and cryosectioned following the protocol for whole brain sections. For detailed procedures see Material and Methods. *preincubation only **main incubation only ***pre- and main incubation only.

| Receptor | [³H]-Ligand | Displacer | Incubation buffer | Preincubation | Main incubation | Rinsing | Change in saturation analysis |
|---|---|---|---|---|---|---|---|
| AMPA | AMPA, 10 nM | Quisqualate, 10 μM | 50 mM Tris-acetate (pH 7.2) + 100 mM KSCN* | 3x10 min at 4°C | 4x4 sec in buffer at 4°C<br>2x2 sec in 2.5% glutaraldehyde in acetone | 1.5-115 nM, slice assay |
| Kainate | Kainate, 1 nM | SYM 2081, 100 μM | 50 mM Tris-citrate (pH 7.1) + 10 mM Calcium acetate | 3x10 min at 4°C | 45 min at 4°C | 3x4 sec at 4°C | 0.1-45 nM, slice assay |
| NMDA | MK 801, 3.3 nM | MK 801, 100 μM | 50 mM Tris-HCl (pH 7.2) + 30 μM Glutamate** + 30 μM Glycine** + 30 μM Spermidine** | 15 min at 4°C | 60 min at 4°C | 2x2 sec in 2.5% glutaraldehyde in acetone<br>2x5 min at 4°C | 0.2-45 nM, slice assay |
| GABA_A | Muscimol, 7.7 nM | GABA, 10 μM | 50 mM Tris-citrate (pH 7.0) | | 40 min at 4°C | 3x3 sec at 4°C<br>1 sec. in a. dest. | 0.5-56 nM, slice assay |
| BZ | Flumazenil, 1 nM | Clonazepam, 2 μM | 170 mM Tris-HCl (pH 7.4) | | 60 min at 4°C | 2x1 min at 4°C<br>1 sec in a. dest. at 4°C | 0.5-16 nM, filtration assay |
| GABA_B | CGP 54626, 2 nM | CGP 55845, 100 μM | 50 mM Tris-HCl (pH 7.2) + 2.5 mM Calcium chloride | | 60 min at 4°C | 3x2 sec at 4°C<br>1 sec in a. dest. at 4°C | 0.1-15 nM, filtration assay |
| A₁ | CPFPX - Cpp(N(tt)p, 4.4 nM | R-PIA, 100 μM | 170 mM Tris-HCl (pH 7.4) + 2 U/ml Adenosine deaminase | | 120 min at 4°C | 2x5 min at 4°C<br>1 sec in a. dest. at 4°C | 0.07-7.0 nM, filtration assay |
| 5-HT₁A | 8-OH-DPAT, 0.3 nM | 5-HT, 1 μM | 170 mM Tris-HCl (pH 7.7) + 10 μM Gpp(NH)p* + 4 mM Calcium chloride | | 60 min at 4°C | 5 min at 4°C<br>3x1 sec in a. dest. | 0.1-5.0 nM, slice assay |
| 5-HT₂ | Ketanserin, 1.14 nM | Mianserin HCl, 10 μM | 170 mM Tris-HCl (pH 7.7) + 0.01% Ascorbate | | 120 min | 2x10 min at 4°C<br>3x1 sec in a. dest. | 0.2-11.0 nM, filtration assay |
| M₁ | Pirenzepine, 10 nM | Pirenzepine, 2 μM | modified Krebs-buffer (pH 7.4): 30.6 mM NaCl, 5.5 mM KCl, 1.2 mM Magnesium sulfate, 1.4 mM Potassium-dihydrogen-phosphate, 5.5 mM D-glucose, 3.2 mM Na-hydrogencarbonate, 2.5 mM Calcium chloride | | 60 min at 4°C | 2x1 min at 4°C<br>3x1 sec in a. dest.<br>1 sec, slice assay | 0.1-45 nM, slice assay |
| M₂ | Oxotremorine-M, 1.7 nM | Carbamoylcholinechloride, 10 μM | 20 mM Hepes-Tris (pH 7.5) + 10 mM Magnesium chloride + 300 nM Pirenzepine*** | 20 min | 60 min | 2x2 min at 4°C<br>1 sec in a. dest. | 0.2-20 nM, filtration assay |
| M₃ | 4-DAMP, 1 nM | Atropine sulfate, 10 μM | 50 mM Tris-HCl (pH 7.4) + 0.1 nM PMSF | 15 min | 45 min | 2x5 min at 4°C<br>1 sec in a. dest. at 4°C | 0.1-10.0 nM, filtration assay |
| nACh | Epibatidine, 0.11 nM | Nicotine, 100 μM | 15 mM HEPES (pH 7.5) + 120 mM NaCl + 5.4 mM KCl + 0.3 mM Magnesium chloride + 1.8 mM Calcium chloride | 20 min | 90 min | 5 min at 4°C<br>1 sec in a. dest. at 4°C | 0.07-4.0 nM, filtration assay |
| D₁ | SCH 23390, 1.67 nM | SKF 83566, 1 μM | 50 mM Tris-HCl (pH 7.4) + 120 mM NaCl + 5 mM KCl + 2 mM Calcium chloride + 1 mM Magnesium chloride + 1 μM Mianserin* | | 90 min | 2x20 min at 4°C<br>1 sec in a. dest. | 0.2-9.0 nM, slice assay |
| α₁ | Prazosin, 0.09 nM | Phentolaminemesylate, 10 μM | modified Sörensen-buffer A: 50 mM D-Na-hydrogenphosphate, B: 50 mM K-dihydrogenphosphate A/B = 4:1 (pH 7.4) | 15 min | 60 min | 5 min<br>1 sec in a. dest. at 4°C | 0.01-3.0 nM, filtration assay |
| α₂ | UK 14,304, 0.64 nM | Phentolaminemesylate, 10 μM | 50 mM Tris-HCl (pH 7.7) + 100 nM Manganese chloride | | 90 min | 5 min<br>1 sec in a. dest. | 0.1-26.0 nM, filtration assay |

## Results

Using quantitative *in-vitro* receptor autoradiography, density measures and laminar distribution patterns of different receptor binding sites were examined in cortical-, subcortical and hippocampal regions of *reeler* and wild type brains. Furthermore, saturation analyses were performed for each receptor subtype on purified membranes of whole brain homogenates to investigate $K_D$ and $B_{max}$ and allow a reliable calculation of regional receptor densities both in mutants and controls (Zilles et al., 2002, 2004).

### $GABA_A$ and $A_1$ receptor densities are decreased in whole brain homogenates of *reeler* mutant mice.

In a first step we performed saturation analyses for all receptor subtypes, to investigate the global effects of the *reelin* mutation on neurotransmitter receptor affinities in wild type and mutant mice brains (Table 2). Since the $K_D$ values are most important when calculating receptor densities from autoradiographs (see below), we also investigated whether the mutation of *reelin* would affect $K_D$ values. No significant changes in $K_D$ values were found between wild type and *reeler* brain homogenates for any of the receptor subtypes investigated. However, we found a significant reduction of the $A_1$ (-23%, $p < 0.001$) and $GABA_A$ (-34%, $p < 0.01$) receptor $B_{max}$ (Fig. 2). Since both receptors in general exhibit high densities in the cerebellum, which is severely reduced in volume in the *reeler* mutant, we performed a further analysis for these receptors exclusively in the cerebellum, where we also found a significant reduction of $A_1$ (-32%, $p < 0.001$) and $GABA_A$ (-61%, $p < 0.001$) receptors.

### Receptor densities exhibit differential regional changes in *reeler* mouse brains

We then asked whether receptor densities were differentially altered in different regions of the brain. Therefore, we investigated the regional receptor densities in Par1, Mot and Put and subregions CA1, CA3, and DG of the hippocampus (Figures 3-6; Table 3).

Although saturation analysis revealed a significant decrease of $GABA_A$ receptor $B_{max}$ in whole brain homogenates (see above), no significant differences were observed in the mean receptor densities in cortical and subcortical brain regions (Table 3). Thus, the overall decrease of $GABA_A$ receptors is restricted mainly to the cerebellum, most likely due to the reduced number of granule cells in the mutant (Mariani et al., 1977). Interestingly, we found a significant reduction of $GABA_A$ related BZ binding sites in the hippocampal CA1 region, as well as in Put (Table 2). Thus, BZ-binding of the $GABA_A$ receptors is reduced in those regions.

In contrast to $GABA_A$ receptor densities, the observed decrease of $A_1$ receptors was reflected in nearly all investigated regions (except CA1) of the brain. Thus, in the *reeler* mutant $A_1$ receptor densities were severely reduced not only in the cerebellum, but also in the forebrain.

Furthermore, in DG the densities of both $GABA_B$ and $5-HT_1$ receptors were significantly reduced in the *reeler* mutant (Table 3). The decrease of $5-HT_1$ receptors was also significant in the hippocampal CA1 region. In contrast, we found a significant increase of $M_2$ receptors in both CA1 and DG (Table 3). In Mot the densities of adrenergic $\alpha_1$ and $\alpha_2$ receptors as well as kainate binding sites were significantly reduced in *reeler* (Table 3).

Taken together, we found differential changes of binding site densities for several receptor subtypes in different brain regions of *reeler* mice, whereas other receptor subtypes remained unaffected by the lack of *reelin*.

## Mutation of *reelin* affects the laminar distribution of receptors in the cerebral cortex in a subclass specific manner.

Since the discovery of an inverted radial distribution of cortical neurons in the *reeler* mutant, it remains still a matter of debate whether the reeler neocortex is indeed "simply" inverted, or whether a laminated structure exists regardless of the malformed *reeler* mouse neocortex.

Since receptors are key elements that link neuronal structure with function, we were interested in the laminar distribution of receptor subtypes in the cortex of wild type and *reeler* mutant mice (Figs. 3-6). Thus, mean profiles were extracted from Par1 of wild type and *reeler* mouse brains (Figs. 7-10) as described in detail in Material and Methods.

The laminar distribution of glutamatergic AMPA, NMDA and kainate receptors within Par1 of the neocortex was generally inverted in the *reeler* mutant (red lines in Fig. 7) when compared with the wild type (black lines in Fig. 7). This was most obvious for kainate receptors (Figs. 3C and 7). While by far the highest kainate receptor densities were found in the deep cortical aspects of the wild type, in *reeler* mutant mice the highest densities were found most superficially near to the pial surface (Figs. 3C and 7). A similar tendency for radial inversion could be observed for the other glutamatergic receptor subtypes investigated, i.e. AMPA (Fig. 3A) and NMDA (Fig. 3B).

The laminar distribution patterns of GABA receptor subtypes were also altered in the *reeler* mutant when compared with the wild type (Figs. 4 and 8). $GABA_A$ receptors exhibited comparatively high densities in the more superficial layers I, II/III and IV in the wild type (compare Figs. 4A and 8, upper panel). In *reeler*, this distribution is disturbed, since deep and superficial aspects of the neocortex show low receptor densities, while they are more or less homogenously distributed throughout the remaining part of the neocortex (Fig. 4A). However, although receptor distribution differs between the mutant and the wild type, a clear inversion of the receptor distribution was not observed. For BZ binding sites this effect became even more obvious (Figs. 4B and 8, middle panel). In the wild type a band of high receptor density was visible at the level of layer IV. Although less distinct, a similar, but more scattered distribution was observed in the *reeler* neocortex (Fig. 4B). A comparison between the mean receptor profiles of *reeler* and wild type Par1 also revealed comparatively similar laminar distribution of BZ binding sites (Fig. 8, middle panel). In contrast to the wild type, the distribution of $GABA_B$ receptors was inverted in the *reeler* mouse neocortex. However, this inversion is less distinct than that observed for the glutamatergic system (compare Fig. 7 with 8, lower panel).

Although mean $A_1$ receptor densities were found to be significantly reduced in *reeler* mouse brains, their laminar distribution pattern in Par1 was quite similar to that of controls (Fig. 5A). $A_1$ receptors were homogenously distributed throughout the cortical depth (Fig. 9, lower panel) in both wild type and reeler brains. Thus, an obvious inverted laminar pattern of this receptor subtype could not be observed.

The analysis of serotonergic 5-HT receptors displayed a very complex pattern. The $5-HT_1$ receptors in general showed relatively high densities in the infragranular layers V and VI in wild type mice. In *reeler* brains, this laminar distribution was severely altered. While the autoradiograph in Fig. 5B (right panel) suggested a more scattered distribution, a closer analysis of mean profiles exhibited a relatively high density of $5-HT_1$ in the superficial aspects of the *reeler* mouse neocortex, hence indicating an inverted distribution in the mutant (Fig. 9, upper panel). Highest $5-HT_2$

receptor densities were found in layers V and VI of the neocortex in the wild type. However, in the mutant this receptor subtype seemed to be randomly scattered throughout the cortical depth (Fig. 9, middle panel). Thus, a clearly inverted receptor distribution could not be defined for this receptor.

A differential expression pattern was observed for the nicotinic (nAch) and muscarinic ($M_1$-$M_3$) acetylcholine receptors (Fig. 6). In the wild type, $M_1$ receptors displayed high densities in superficial layer II/III, while layers V and VI were characterized by comparatively low densities. In the *reeler* mutant this pattern was basically inverted, but additionally the receptors were more scattered throughout cortical depth. Concerning $M_2$ receptors in the wild type high densities were found in layer IV, while expression in layers I-III and V was comparatively low. In the *reeler* cortex this band of high density was virtually absent, but receptors were scattered throughout cortical depth. Thus, $M_2$ receptors did not show a clear laminar inversion in the *reeler* cortex but layer IV synaptic connections (i.e. these neurons receive the thalamocortical input) seemed to be more affected than other layers. The laminar distribution of the $M_3$ receptor subtype resembled that of $M_1$ receptors (compare Fig. 6A with 6C), i.e. highest densities were present in the superficial layers of the wild type. The laminar M3 receptor distribution was clearly inverted in *reeler* mice, and appears diluted over the infragranular layers.

Very low densities of the nAch receptor subtype were observed in the neocortex of both wild type and *reeler*. Relatively high densities were found in the deeper cortical layers of wild type brains whereas comparatively high densities were observed in the more superficial parts of the neocortex in the *reeler* mutant (Figs. 6D and 10), thus indicating a laminar inversion in the *reeler* cortex.

The distribution patterns of adrenergic $\alpha_1$ and $\alpha_2$ receptors exhibited an altered distribution in the mutant neocortex when compared to that of the wild type (Supplementary Fig. 1). However, a clear inversion of the laminar distribution of these receptors could not be observed.

**Table 2: Dissociation constant ($K_D$) and maximum binding capacity ($B_{max}$) of different neurotransmitter receptors in wild type and reeler brains.** Saturation analyses were performed using purified membrane preparations of whole brain homogenates as described in Material and Methods. Binding conditions are summarized in Table 1. In reeler mice the $B_{max}$ of $A_1$ and $GABA_A$ receptors were significantly decreased, while the $K_D$ values of all investigated receptor subtypes were unchanged. Significant differences are highlighted in bold text and marked by one (p ≤ 0.01) or two (p ≤ 0.001) asterisks, respectively.

| Receptor | $K_D$ WT | $K_D$ Rn | $B_{max}$ WT | $B_{max}$ Rn |
|---|---|---|---|---|
| AMPA | 15.9 ± 2.5 | 15.1 ± 1.3 | 922 ± 91 | 1066 ± 164 |
| Kainate | 4.1 ± 0.5 | 4.0 ± 0.5 | 710 ± 46 | 629 ± 41 |
| NMDA | 14.5 ± 4.8 | 14.2 ± 2.9 | 217 ± 51 | 202 ± 32 |
| $A_1$ | 0.4 ± 0.03 | 0.3 ± 0.04 | 658 ± 31 | **\*\*509 ± 21** |
| $GABA_A$ | 7.1 ± 0.7 | 6.8 ± 0.5 | 1856 ± 96 | **\*1228 ± 165** |
| BZ | 2.3 ± 0.1 | 2.3 ± 0.1 | 1389 ± 27 | 1364 ± 16 |
| $GABA_B$ | 1.3 ± 0.1 | 1.5 ± 0.2 | 423 ± 11 | 378 ± 21 |
| $5-HT_1$ | 0.4 ± 0.0 | 0.4 ± 0.1 | 177 ± 10 | 177 ± 6 |
| $5-HT_2$ | 3.5 ± 0.8 | 2.2 ± 0.7 | 455 ± 75 | 320 ± 73 |
| $D_1$ | 0.7 ± 0.1 | 0.6 ± 0.0 | 668 ± 61 | 622 ± 22 |
| $M_1$ | 18.4 ± 3.9 | 23.7 ± 2.3 | 188 ± 22 | 131 ± 9 |
| $M_2$ | 2.9 ± 0.2 | 3.8 ± 0.7 | 226 ± 21 | 250 ± 36 |
| $M_3$ | 1.9 ± 0.2 | 2.0 ± 0.1 | 1013 ± 67 | 859 ± 60 |
| nAch | 0.04 ± 0.003 | 0.03 ± 0.002 | 22 ± 1 | 21 ± 1 |
| $\alpha_1$ | 0.1 ± 0.04 | 0.2 ± 0.03 | 340 ± 61 | 383 ± 44 |
| $\alpha_2$ | 2.2 ± 0.4 | 2.3 ± 0.5 | 150 ± 9 | 82 ± 6 |

**Fig. 2: $A_1$ and $GABA_A$ receptor densities are significantly reduced in *reeler* whole brain homogenates while receptor affinities remain unchanged.** Whole brain membrane preparations, as exemplified for $A_1$ and $GABA_A$, were incubated with increasing concentrations of tritiated ligand and specific binding was assessed. Left upper panel Saturation analysis using non-linear regression showing a significant reduction in $A_1$ receptor binding ($B_{max}$) in *reeler* brain tissue, while receptor affinity remains unchanged. Right upper panels: Scatchard plots of the respective data. Left lower panel, saturation analysis using non-linear regression showing a significant reduction in $GABA_A$ receptor binding ($B_{max}$) in *reeler* brain tissue, while receptor affinity remains unchanged. Right lower panels: Scatchard plots of the respective data. For all saturation analyses non-specific binding was less than 5%. B, binding; F, ligand concentration.

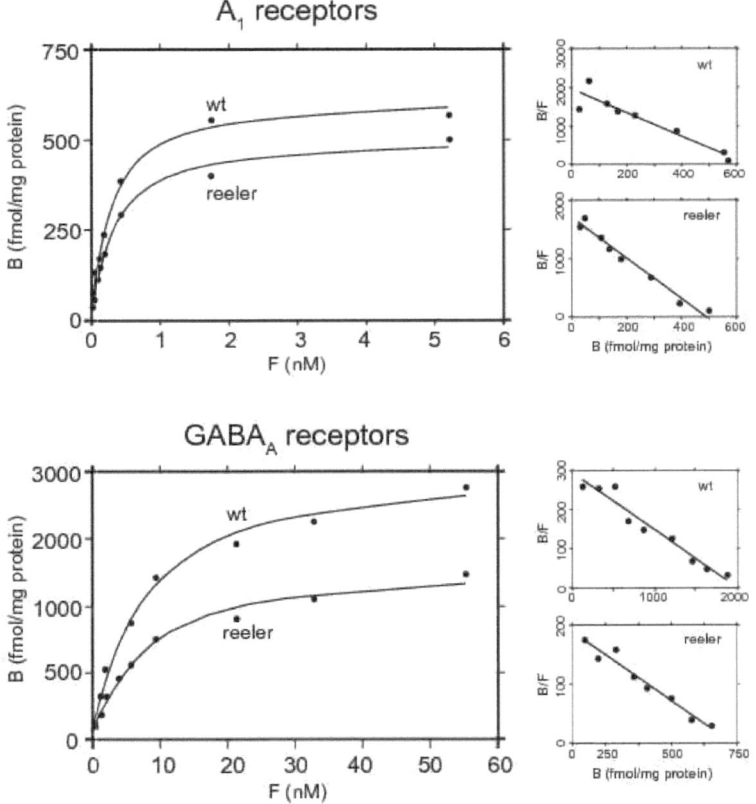

Table 3: Neurotransmitter receptor densities in different regions of wild type and reeler mice brains. Receptor densities were investigated using quantitative in vitro receptor autoradiography as described in Material and Methods. Binding conditions are summarized in Table 1. In reeler brains several neurotransmitter receptor densities were altered when compared to the wild type. While most changes were regionally restricted, $A_1$ receptors were significantly decreased in all regions except CA1. Significant differences are highlighted in bold text and marked by one (p ≤ 0.05) or two (p ≤ 0.01) asterisks, respectively. Receptor densities were calculated for the somatosensory cortex (Par 1), motor cortex (Mot), putamen (Put) and the subregions CA1, CA3, and dentate gyrus (DG) of the hippocampus.

| | CA1 WT | CA1 Rn | CA3 WT | CA3 Rn | DG WT | DG Rn | Put WT | Put Rn | Par1 WT | Par1 Rn | Mot WT | Mot Rn |
|---|---|---|---|---|---|---|---|---|---|---|---|---|
| AMPA | 3219 ± 93 | 3116 ± 55 | 2521 ± 52 | 2551 ± 43 | 2638 ± 47 | 2413 ± 43 | 1328 ± 28 | 1434 ± 36 | 1383 ± 34 | 1672 ± 115 | 1519 ± 107 |
| Kainate | 515 ± 9 | 477 ± 5 | 927 ± 11 | 966 ± 14 | 1027 ± 11 | 978 ± 14 | 1351 ± 22 | 1395 ± 24 | 951 ± 22 | 813 ± 16 | 1059 ± 35 | **861 ± 17 |
| NMDA | 3182 ± 27 | 2907 ± 52 | 1643 ± 28 | 1656 ± 33 | 2295 ± 30 | 2037 ± 33 | 1084 ± 15 | 1050 ± 22 | 1159 ± 49 | 998 ± 55 | 1203 ± 71 | 1007 ± 128 |
| $A_1$ | 2834 ± 38 | 2565 ± 28 | 2320 ± 31 | ***2030 ± 18 | 1925 ± 25 | ***1744 ± 14 | 1545 ± 16 | **1384 ± 8 | 2055 ± 19 | ***1706 ± 17 | 1600 ± 26 | ***1335 ± 25 |
| $GABA_A$ | 1951 ± 24 | 1826 ± 48 | 1070 ± 19 | 1140 ± 36 | 2523 ± 33 | 2331 ± 55 | 1694 ± 28 | 1722 ± 31 | 2477 ± 73 | 2328 ± 127 | 2018 ± 73 | 2046 ± 237 |
| BZ | 3371 ± 56 | **2785 ± 39 | 2281 ± 37 | 2277 ± 42 | 3302 ± 75 | 2819 ± 33 | 1283 ± 15 | **1046 ± 21 | 3180 ± 79 | 2869 ± 41 | 2901 ± 96 | 2504 ± 72 |
| $GABA_B$ | 4042 ± 17 | 4208 ± 24 | 4743 ± 35 | 4954 ± 28 | 4555 ± 38 | **4069 ± 26 | 2247 ± 28 | **1046 ± 21 | 3961 ± 42 | 3845 ± 54 | 4540 ± 67 | 4331 ± 152 |
| $5-HT_1$ | 611 ± 11 | *494 ± 13 | 86 ± 3 | 86 ± 5 | 198 ± 4 | *136 ± 7 | n.m. | 99 ± 9 | 74 ± 4 | 105 ± 12 | 89 ± 9 |
| $5-HT_2$ | 178 ± 9 | 203 ± 5 | 159 ± 7 | 198 ± 6 | 164 ± 15 | 173 ± 4 | 554 ± 11 | 479 ± 8 | 256 ± 35 | 245 ± 5 | 268 ± 31 | 186 ± 14 |
| $D_1$ | n.m. | n.m. | n.m. | n.m. | n.m. | n.m. | 1039 ± 37 | 1021 ± 14 | n.m. | n.m. | n.m. | n.m. |
| $M_1$ | 1753 ± 48 | 1562 ± 32 | 1054 ± 27 | 1098 ± 24 | 1627 ± 44 | 1424 ± 38 | 1591 ± 38 | 1472 ± 38 | 1013 ± 28 | 732 ± 73 | 1144 ± 70 | 1005 ± 101 |
| $M_2$ | 273 ± 6 | **366 ± 6 | 427 ± 12 | 516 ± 11 | 202 ± 4 | *245 ± 5 | 454 ± 8 | 416 ± 9 | 824 ± 22 | 821 ± 21 | 632 ± 37 | 549 ± 52 |
| $M_3$ | 1415 ± 13 | 1409 ± 17 | 1020 ± 11 | 1067 ± 20 | 1243 ± 13 | 1224 ± 23 | 1367 ± 21 | 1344 ± 25 | 1226 ± 16 | 1069 ± 58 | 1112 ± 29 | 1078 ± 130 |
| nAch | 32 ± 5 | 24 ± 3 | 14 ± 5 | 12 ± 1 | 47 ± 9 | 42 ± 4 | 182 ± 11 | 147 ± 7 | 137 ± 12 | 123 ± 9 | 166 ± 10 | 133 ± 13 |
| $\alpha_1$ | 146 ± 3 | 144 ± 2 | 141 ± 3 | 132 ± 3 | 119 ± 4 | 123 ± 1 | 103 ± 4 | 106 ± 4 | 342 ± 10 | 353 ± 15 | 552 ± 25 | *428 ± 33 |
| $\alpha_2$ | n.m. | n.m. | n.m. | n.m. | 140 ± 4 | 141 ± 5 | 98 ± 3 | 76 ± 2 | 145 ± 10 | 104 ± 6 | 139 ± 16 | *83 ± 14 |

**Fig. 3: Distribution of glutamate receptor subtypes is altered in *reeler* mutant mice.** .Color coded autoradiographic images of a wild type (left) and *reeler* mouse (right) hemisphere visualizing the regional and laminar receptor distribution patterns. Hemispheres were color-coded using different scaling for optimal visualization of receptor density pattern. In the color scales below each image receptor density is given as fmol/mg protein. For details of regional receptor densities see Table 3 and results. A AMPA receptors; B NMDA receptors; C kainate receptors.

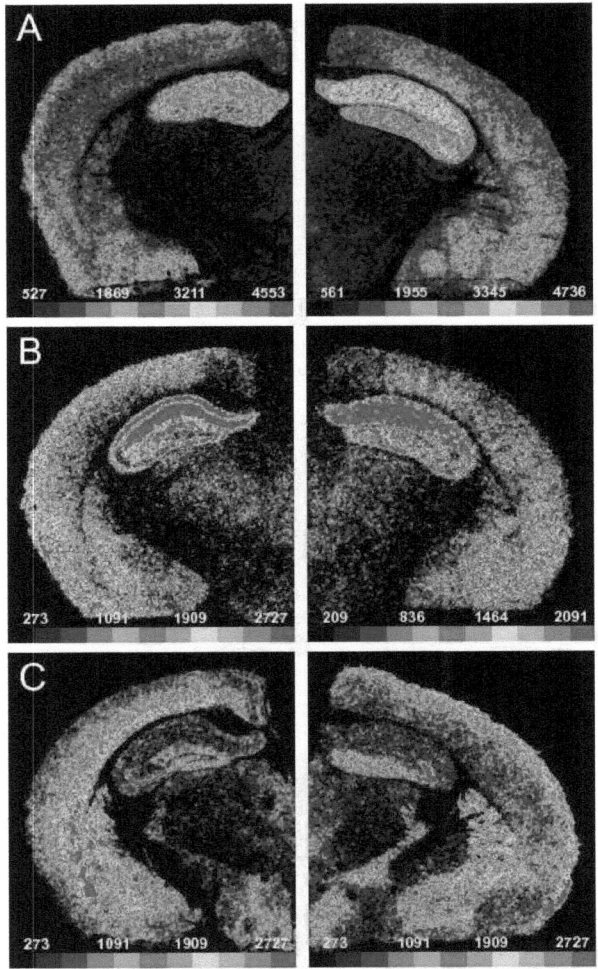

Reeler mice

**Fig. 4: Distribution of GABA receptor subtypes in wild type and *reeler* mice.** Color coded autoradiographs of a representative section of a wild type (left) and *reeler* mouse (right) hemisphere showing the distribution of $GABA_A$ receptors (A), benzodiazepine binding sites (B) and $GABA_B$ receptors (C). Note the different color scales below each image. Receptor density is given as fmol/mg protein. For details of regional receptor densities see Table 3 and results.

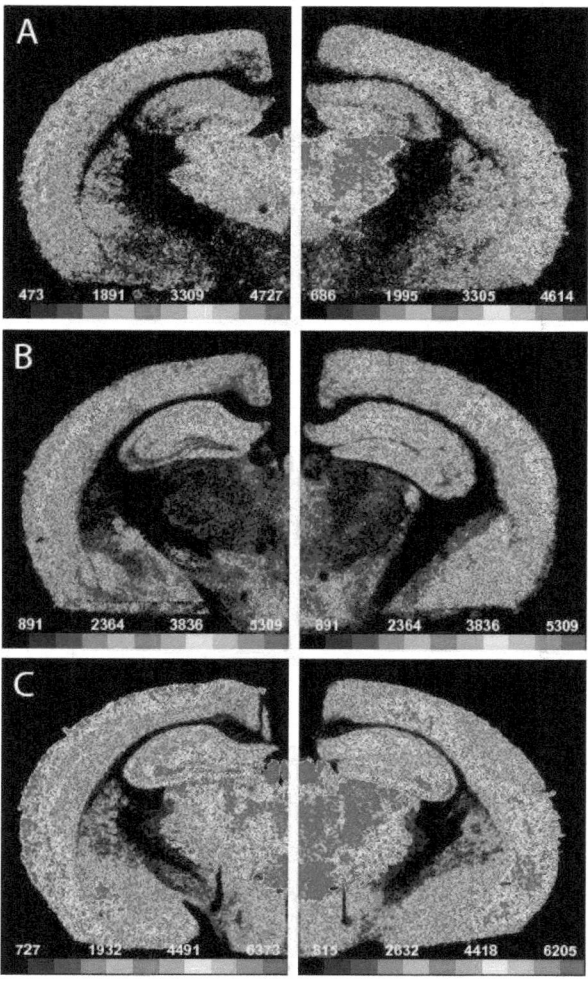

119

**Fig. 5: Distribution of A1 and serotonergic receptor subtypes in wild type and *reeler* mice.** Color coded autoradiographs of a representative section of a wild type (left) and *reeler* mouse (right) hemisphere showing the distribution of A1 (A), 5-$HT_1$ (B), and 5-$HT_2$ receptors (C). Note the different color scales below each image. Receptor density is given as fmol/mg protein. For details of regional receptor densities see Table 3 and results.

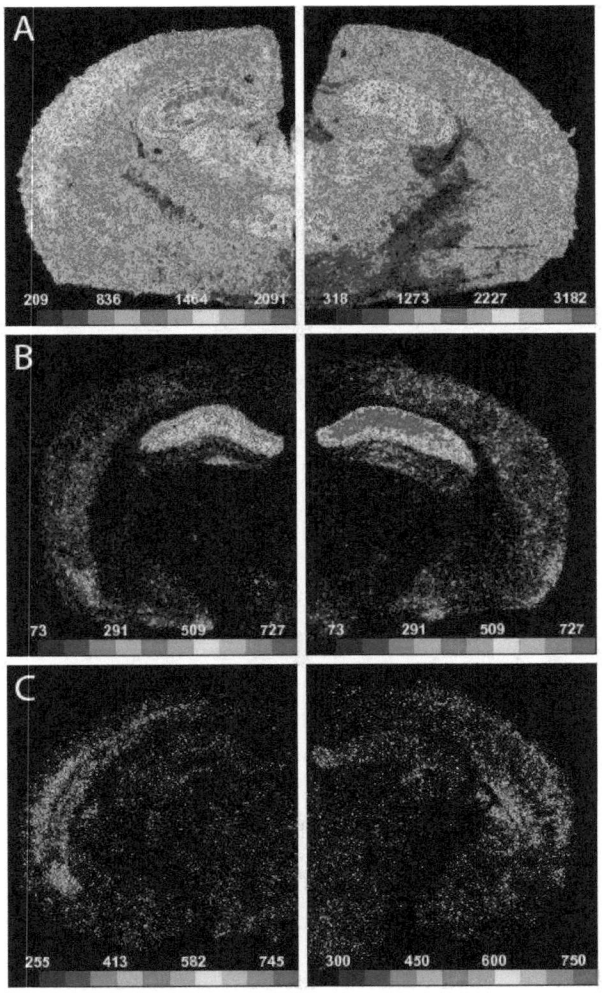

**Fig. 6: Muscarinic and nicotinic acetylcholine receptors show different distribution patterns in wild type and *reeler* mice.** Color coded autoradiographs of a representative section of a wild type (left) and *reeler* mouse (right) hemisphere showing the distribution of muscarinic $M_1$ (A), $M_2$ (B), $M_3$ (C) and nicotinic acetylcholine receptors (D). Note the different color scales below each image. Receptor density is given as fmol/mg protein. For details of regional receptor densities see Table 3 and results.

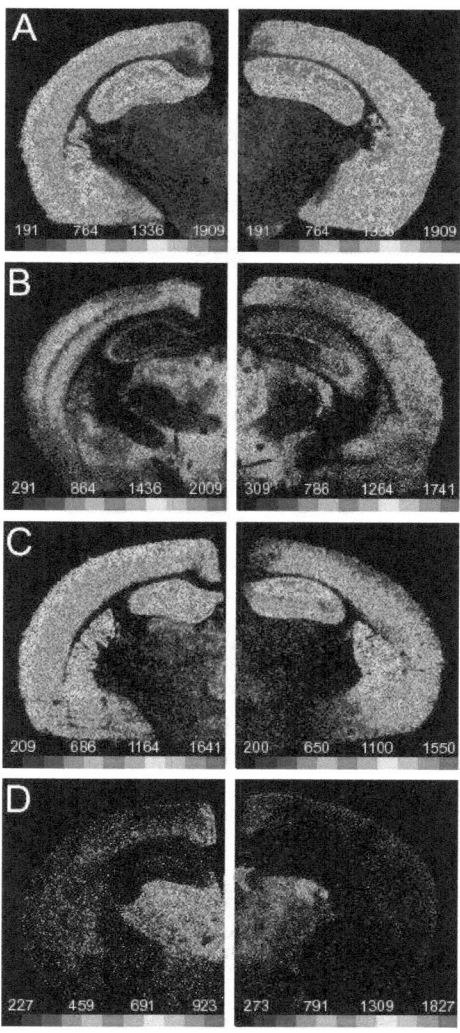

**Fig. 7: Laminar distribution of glutamatergic receptors.** Laminar profiles through the cortical depth for AMPA (upper panel), NMDA (middle panel) and kainate (lower panel) receptors. Density profiles (± SEM) were extracted and averaged in the primary somatosensory cortex of five wild type (black line) and five *reeler* brains (red line) as described in Material and Methods. The total cortical depth from the pial surface to the white matter border was normalized to 100%, while the area under the curve was normalized to 1.

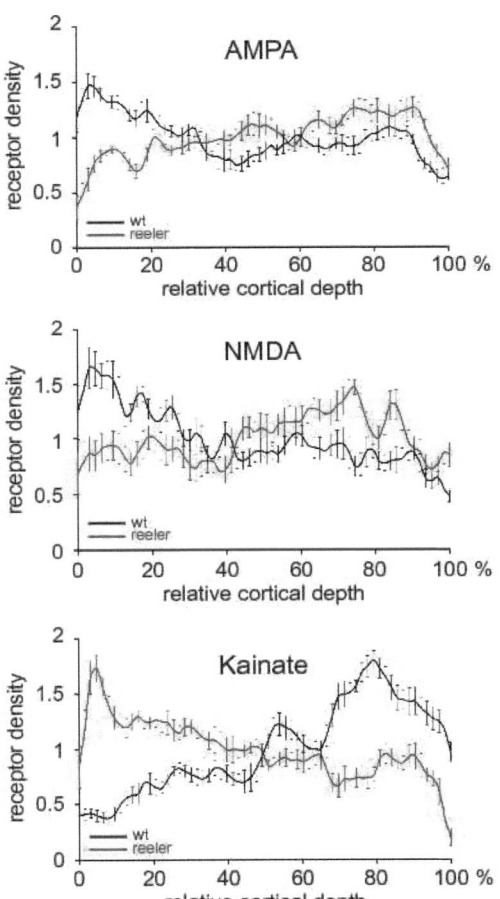

**Fig. 8: Laminar distribution of GABA receptors.** Laminar profiles of receptor densities (± SEM) through the cortical depth for $GABA_A$ (upper panel) receptors, benzodiazepine binding sites (middle panel) and $GABA_B$ (lower panel) receptors in the primary somatosensory cortex of wild type (black line) and *reeler* brains (red line). While differences in the laminar distribution pattern exist for $GABA_A$ and $GABA_B$ receptors more subtle changes were observed for benzodiazepine binding sites. For normalization see legend Fig. 7.

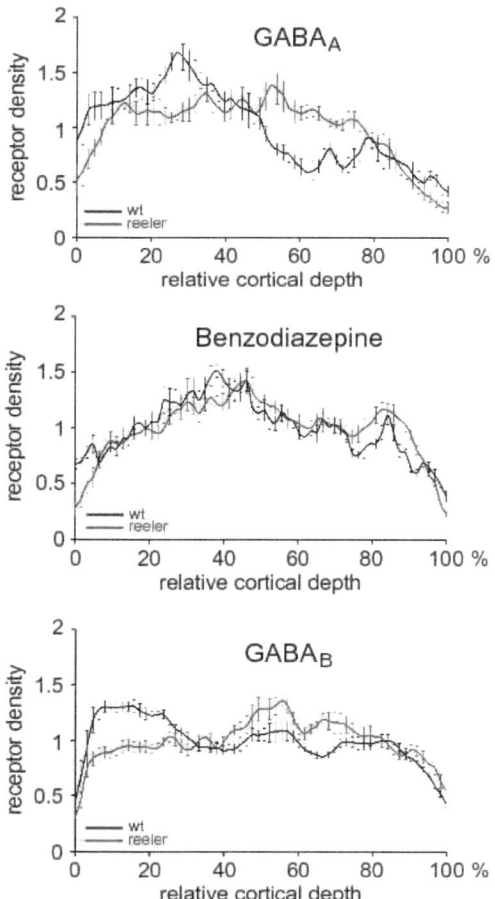

**Fig. 9: Laminar distribution of $A_1$ and serotonergic receptors.** Laminar profiles of receptor densities (± SEM) through the cortical depth for $A_1$ (upper panel), $5\text{-}HT_1$ (middle panel) and $5\text{-}HT_2$ (lower panel) receptors in the primary somatosensory cortex of wild type (black line) and *reeler* brains (red line). For normalization see legend Fig. 7.

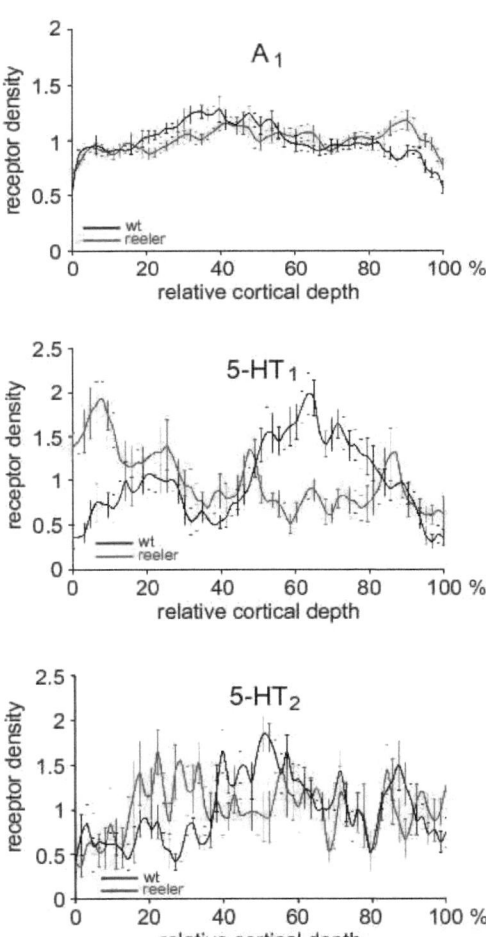

**Fig. 10: Laminar distribution of muscarinic and nicotinic acetylcholine receptors.** Laminar profiles of receptor densities (± SEM) through the cortical depth for muscarinic $M_1$ (upper panel), $M_2$ (upper middle panel), $M_3$ (lower middle panel) and nicotinic (lower panel) acetylcholine receptors in the primary somatosensory cortex of wild type (black line) and *reeler* brains (red line). Whereas muscarinic $M_1$ and $M_3$ receptors show an inversion in their laminar distribution, no inversion was observed for the muscarinic $M_2$ and the nicotinic acetylcholine receptors. For normalization see legend Fig. 7.

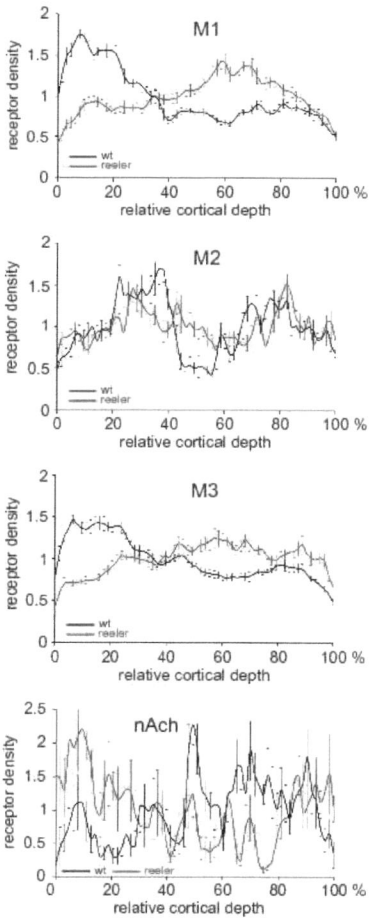

## Discussion

The regional densities of sixteen different neurotransmitter receptors were analyzed for the first time in neo- and allocortical areas as well as in subregions of the hippocampus of *reeler* mice and compared with controls. Furthermore, laminar profiles of receptor density were extracted from Par1, indicating that most receptors exhibit a clear laminar inversion in the *reeler* brain (i.e. AMPA, kainate, NMDA, $GABA_B$, $5-HT_1$, $M_1$, $M_3$, nAch), while others did not demonstrate an obvious inversion ($A_1$, $GABA_A$, BZ, $5-HT_2$, $M_2$, $\alpha_1$, $\alpha_2$). Additionally, our data demonstrated significant changes of whole brain $B_{max}$ for $A_1$ and $GABA_A$ receptors in *reeler* brains. While the decrease of $GABA_A$ receptors proved to be restricted to the cerebellum, other receptor subtypes ($A_1$, kainate, BZ binding sites, $5-HT_1$, $M_2$, $\alpha_1$ and $\alpha_2$) showed regional changes of their density in the forebrain.

## Laminar distribution of receptor subtypes

The mammalian neocortex consists of six layers of neurons with distinct functional and morphological characteristics. The layers are generated during ontogeny in an inside-out sequence, with early born neurons located in deep and late born neurons in superficial layers (Angevine and Sidman, 1961; Berry and Rogers, 1965; Rakic and Caviness, 1995; Takahashi et al., 1999). The extracellular matrix protein *reelin* is critical for the migration of neurons to their final laminar position acting via two receptors: the very low density lipoprotein receptor (VLDLR) and the apolipoprotein E receptor (ApoER2). The proper migration of late generated neurons, requires ApoER2 whereas VLDLR mediates a *reelin* caused stop signal that prevents neurons migrating into the cell-poor marginal zone (Hack et al., 2007). Mutations in the *reelin*, the Dab1 and in both the VLDLR and ApoER2 genes result in a *reeler*-like phenotype characterized by a severely altered cortical layering despite a virtually normal development of the preplate (Caviness, 1982; Howell et al., 1997; Sheldon et al., 1997; Sheppard and Pearlmann, 1997; Ware et al., 1997; Trommsdorff et al., 1999; for review see Lambert de Rouvroit and Goffinet, 1998).

We hypothesized that the laminar receptor distribution pattern in Par1 should basically mirror this inverted cellular organization. The laminar distribution of AMPA, kainate, NMDA, $GABA_B$, $M_1$, $M_3$, nAch and $5-HT_1$ receptor densities reflected this assumption, while other receptor subtypes ($GABA_A$, BZ, $M_2$, $5-HT_{2A}$, $\alpha_1$, $\alpha_2$ and $A_1$) did not show a clearly inverted distribution pattern (Figs. 7-10, Suppl. Fig. 1). These observations are partially confirmed by Dekimoto et al. (2010) who used *in situ* hybridization to study the distribution of layer specific mRNAs in *reeler* and control brains. Similar to our findings, they reported that layer specificity in the *reeler* somatosensory cortex exhibits a complex pattern ranging from splitting of layers (layer VI, characterized by Tbr1 expression) to a complete disruption of layer integrity and dispersion of the respective neurons (layer V, characterized by ER81 expression). None of the investigated layer specific markers demonstrated a clear radial inversion. However, labeling of layer specific markers by *in situ* hybridization is restricted to the cellular level. It has to be taken into account that neurotransmitter receptors distribution is cell and target specific for a given neuronal cell type as shown at the subcellular level. While a given receptor may be localized at the dendrite, the respective soma can be localized in a completely different layer. Additionally, some neurotransmitter receptors are more or less homogenously distributed throughout the cortical depth in the wild type (e.g. $A_1$ receptors, cf. figures 5A and 9), and thus, it is not possible to state whether their distribution is inverted in the *reeler* brain.

In addition, neurotransmitter receptors, in particular glutamate and GABA receptors are involved in the migration (Behar et al., 1999; Lopez-Bendito et al., 2003) and positioning of neurons in their appropriate target layer (for review see Lujan et al., 2005). Lack of *reelin* leads to an incorrect migration and false positioning of neurons and as a consequence may also result in the differential expression and different cell-specific subcellular distribution of neurotransmitter receptor subtypes in correctly versus malpositioned late born neurons. Thus, it can be assumed that the establishment and maintenance of the different neurotransmitter systems may be disturbed in *reeler* mice during embryonic and early postnatal development, and hence directly affect the expression of specific neurotransmitter receptor subtypes.

## *Reelin* in synaptic transmission and plasticity

Beside its known function during cortical development, recent studies have focused on the role of *reelin* in the adult brain. Two major functions seem to be associated with the *reelin* pathway. First, recent studies indicate *reelin* as an important factor to preserve hippocampal lamination, i.e. application of recombinant *reelin* in a kainate model of epilepsy inhibits seizure-associated granule cell dispersion of the dentate gyrus (Müller et al., 2009). Vice versa, inhibition of *reelin* by hippocampal injection of the CR-50 antibody results in massive granule cell dispersion (Heinrich et al., 2006).

Second, *reelin* regulates synaptic plasticity, e.g. by increasing the surface mobility of the NR2B subunit of the NMDA receptor (Groc et al., 2007), modulation of receptor subunit composition and activity of NMDA and AMPA receptors (Chen et al., 2005; Sinagra et al., 2005; Qiu et al., 2006) and by enhancing long-term potentiation (Marrone et al., 2006; Beffert et al., 2004, 2005; Weeber et al., 2002). Thus, it was hypothesized that *reelin* deficiency may influence neurotransmitter receptor densities.

It was suggested before that *reelin* is involved in synaptic plasticity by regulating neurotransmitter receptor expression. Using saturation analysis Matsuzaki et al. (2007) demonstrated that $D_1$ and $D_2$ receptor binding is decreased in striatal homogenates of *reeler* mice. Furthermore, dopamine receptor dependent methamphetamine-induced hyperlocomotion is reduced in wild type mice after in vivo inhibition of *reelin* function. Thus, these results indicate a role for *reelin* in the regulation of dopaminergic neurotransmission.

In contrast, we did not find significant changes of $D_1$ receptor expression in *reeler* mice. While their methodological approach was similar to ours, Matsuzaki and colleagues (2007) used a different strain of reeler mice (Reeler$^{rl-j}$, Jackson Laboratories) bearing a different allele of the reeler mutation. In Reeler$^{Orl}$ mice, used in the present study, a truncated *reelin* protein is produced but not secreted (Bergeyck et al., 1997), while in Reeler$^{rl-j}$ mice a deletion of approximately 150 kb 3' sequences leads to a lack of functional *reelin* (D'Arcangelo et al., 1996). Thus, different alleles of the *reeler* mutation may lead to different effects on dopamine receptor expression.

Here, significant changes in mean densities of several neurotransmitter receptors were found in *reeler* mice, whereas others remained unaffected. In the *reeler* somatosensory cortex significant differences could only be found for the $A_1$ receptors. This result may support earlier findings that the thalamocortical input to the barrel field is generally maintained in *reeler*, although the barrels are structurally altered (reviewed by Lopez-Bendito and Molnar, 2003). This assumption is supported by previous cytoarchitectural and electrophysiological studies that proposed a general preservation of area specific characteristics in this mutant (Simmons et al., 1982; Caviness and Frost, 1983; Simmons and Pearlman, 1983; Terashima et al., 1987; Strazielle et al., 2006).

## Implications for *reelin* function in neurodegenerative diseases

The *reelin* pathway has been suggested to be involved in the pathology of very different neurological diseases like lissencephaly, schizophrenia, epilepsy and Alzheimer's disease (reviewed by D'Arcangelo et al., 2006; Deutsch et al., 2006; Fatemi, 2008; Haas and Frotscher, 2010).

Here, we describe a massive decrease of $A_1$ receptor densities in the *reeler* mutant. Activation of $A_1$ receptors results in a reduction of synaptic glutamate release (Prestwich et al., 1987), thereby regulating neuronal activity, synaptic plasticity and learning (reviewed by Fredholm et al., 2005; Stone et al., 2009). The importance of $A_1$ receptors for synaptic transmission is underlined by the fact that a significant decline of $A_1$ receptor density was numerously demonstrated in brains of patients suffering from Alzheimer's disease (AD) (Jansen et al., 1990; Kalaria et al., 1990; Jaarsma et al., 1991; Ikeda et al., 1993; Ulas et al., 1993; Fukumitsu et al., 2008). An increasing body of evidence also implies a role for the *reelin* pathway in AD. *Reelin* signaling antagonizes beta-amyloid-induced suppression of NMDA receptor-mediated long-term potentiation (Durakoglugil et al., 2009), and *reelin* is found in plaques of ß-amyloid precursor protein (APP) and presenilin-1 double-transgenic mice (Wirths et al., 2001). Furthermore, *reelin* expression is decreased in the entorhinal cortex of human APP transgenic mice as well as in human AD patients (Chin et al., 2007). In addition, the neuronal expression, glycosylation pattern and cerebrospinal fluid levels of *reelin* were shown to be differentially altered in human AD (Saéz-Valero et al., 2003; Botella-López et al., 2006). The *reeler* mouse is admittedly not an established model for the complex mechanism of human AD. However, these correlations make it tempting to speculate that *reelin* mediated changes of neurotransmitter receptor expression might possibly link the *reelin* pathway and the molecular mechanisms of AD.

In summary, our results on the density and distribution pattern of neurotransmitter receptors in *reeler* mice point to a role of *reelin* in the regulation of different neurotransmitter systems thereby directly or indirectly controlling synaptic transmission and plasticity.

**Supplementary Fig. 1: Distribution of adrenergic $\alpha_1$ and $\alpha_2$ receptor subtypes in wild type and reeler mice.**

Color coded autoradiographs of a representative section of a wild type (left) and *reeler* mouse (right) hemisphere showing the distribution of $\alpha_1$ (A, A1) and $\alpha_2$ (B, B1) receptors. A2, B2, Laminar distribution of $\alpha_1$ (upper right panel) and $\alpha_2$ (lower right panel) receptors in the primary somatosensory cortex of wild type and the *reeler* mutant mice.

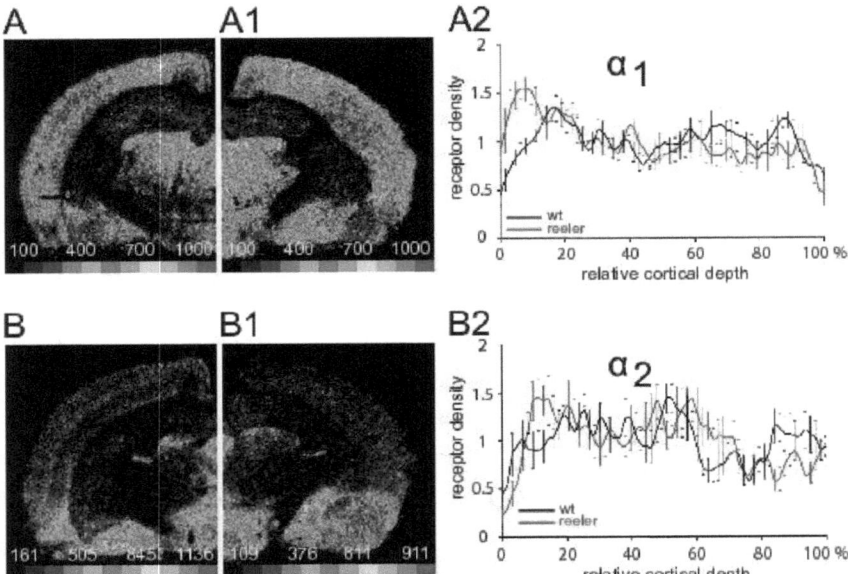

## References

Angevine JB, Sidman RL (1961) Autoradiographic study of cell migration during histogenesis of cerebral cortex in the mouse. Nature 192:766-768.

Beffert U, Weeber EJ, Morfini G, Ko J, Brady ST, Tsai LH, Sweatt JD, Herz J (2004) Reelin and cyclin-dependent kinase 5-dependent signals cooperate in regulating neuronal migration and synaptic transmission. J Neurosci 24:1897-1906.

Beffert U, Weeber EJ, Durudas A, Qiu S, Masiulis I, Sweatt JD, Li WP, Adelmann G, Frotscher M, Hammer RE, Herz J. (2005) Modulation of synaptic plasticity and memory by Reelin involves differential splicing of the lipoprotein receptor Apoer2. Neuron 47:567-79.

Behar TN, Scott CA, Greene CL, Wen X, Smith SV, Maric D, Liu QY, Colton CA, Barker JL (1999) Glutamate acting at NMDA receptors stimulates embryonic cortical neuronal migration. J Neurosci 19:4449-4461.

de Bergeyck V, Nakajima K, Lambert de Rouvroit C, Naerhuyzen B, Goffinet AM, Miyata T, Ogawa M, Mikoshiba K. (1997) A truncated Reelin protein is produced but not secreted in the 'Orleans' reeler mutation (Reln[rl-Orl]). Brain Res Mol Brain Res 50:85-90.

Berry M, Rogers AW (1965) The migration of neuroblasts in the developing cerebral cortex. J Anat 99:691-709.

Botella-Lopez A, Burgaya F, Gavin R, Garcia-Ayllon MS, Gomez-Tortosa E, Pena-Casanova J, Urena JM, Del Rio JA, Blesa R, Soriano E, Saéz-Valero J (2006) Reelin expression and glycosylation patterns are altered in Alzheimer's disease. Proc Natl Acad Sci U S A 103:5573-5578.

Caviness VS, Jr. (1982) Neocortical histogenesis in normal and reeler mice: a developmental study based upon [$^3$H]thymidine autoradiography. Brain Res 256:293-302.

Caviness VS, Frost DO (1983) Thalamocortical projections in the reeler mutant mouse. J Comp Neurol 219:182-202.

Chen Y, Beffert U, Ertunc M, Tang TS, Kavalali ET, Bezprozvanny I, Herz J (2005) Reelin modulates NMDA receptor activity in cortical neurons. J Neurosci 25:8209-8216.

Chin J, Massaro CM, Palop JJ, Thwin MT, Yu GQ, Bien-Ly N, Bender A, Mucke L (2007) Reelin depletion in the entorhinal cortex of human amyloid precursor protein transgenic mice and humans with Alzheimer's disease. J Neurosci 27:2727-2733.

Cremer CM, Palomero-Gallagher N, Bidmon HJ, Schleicher A, Speckmann EJ, Zilles K (2009) Pentylenetetrazole-induced seizures affect binding site densities for GABA, glutamate and adenosine receptors in the rat brain. Neuroscience 163:490-499.

Cremer CM, Bidmon H-J, Görg B, Palomero-Gallagher N, Lopez Escobar J, Speckmann E-J, Zilles, K. (2010) Inhibition of glutamate/glutamine cycle in vivo results in decreased benzodiazepine binding and differentially regulated GABAergic subunit expression in the rat brain. Epilepsia, in press.

D'Arcangelo G (2005) The reeler mouse: anatomy of a mutant. Int Rev Neurobiol 71:383-417.

D'Arcangelo G. (2006) Reelin mouse mutants as models of cortical development disorders. Epilepsy Behav 8:81-90.

D'Arcangelo G, Miao GG, Curran T. (1996) Detection of the reelin breakpoint in reeler mice. Brain Res Mol Brain Res 39:234-236.

Dekimoto H, Terashima T, Katsuyama Y. (2010) Dispersion of the neurons expressing layer specific markers in the reeler brain. Dev Growth Differ 52:181-93.

Deutsch SI, Rosse RB, Lakshman RM. (2006) Dysregulation of tau phosphorylation is a hypothesized point of convergence in the pathogenesis of alzheimer's disease, frontotemporal dementia and schizophrenia with therapeutic implications. Prog Neuropsychopharmacol Biol Psychiatry 30:1369-80.

Durakoglugil MS, Chen Y, White CL, Kavalali ET, Herz J (2009) Reelin signaling antagonizes beta-amyloid at the synapse. Proc Natl Acad Sci U S A 106:15938-15943.

Fatemi SH, editor (2008) Reelin Glycoprotein: Structure, Biology and Roles in Health and Disease. Springer Science+Business Media, New York, USA

Fredholm BB, Chen JF, Cunha RA, Svenningsson P, Vaugeois JM (2005a) Adenosine and brain function. Int Rev Neurobiol 63:191-270.

Fredholm BB, Chen JF, Masino SA, Vaugeois JM (2005b) Actions of adenosine at its receptors in the CNS: insights from knockouts and drugs. Annu Rev Pharmacol Toxicol 45:385-412.

Fukumitsu N, Ishii K, Kimura Y, Oda K, Hashimoto M, Suzuki M, Ishiwata K. (2008) Adenosine A(1) receptors using 8-dicyclopropylmethyl-1-[(11)C]methyl-3-propylxanthine PET in Alzheimer's disease. Ann Nucl Med 22:841-847.

Groc L, Choquet D, Stephenson FA, Verrier D, Manzoni OJ, Chavis P. (2007) NMDA receptor surface trafficking and synaptic subunit composition are developmentally regulated by the extracellular matrix protein Reelin. J Neurosci 27:10165-75.

Haas CA, Frotscher M (2010) Reelin deficiency causes granule cell dispersion in epilepsy. Exp Brain Res 200:141-9.

Hack I, Hellwig S, Junghans D, Brunne B, Bock HH, Zhao S, Frotscher M (2007) Divergent roles of ApoER2 and Vldlr in the migration of cortical neurons. Development 134:3883-3891.

Hamburgh M (1960) Observations on the neuropathology of "Reeler", a neurological mutation in mice. Experientia 16:460-461.

Hamburgh M (1963) Analysis of the Postnatal Developmental Effects of "Reeler," a Neurological Mutation in Mice. A Study in Developmental Genetics. Dev Biol 19:165-185.

Heinrich C, Nitta N, Flubacher A, Müller MC, Fahrner A, Kirsch M, Freiman T, Suzuki F, Depaulis A, Frotscher M, Haas CA (2006) Reelin deficiency and displacement of mature neurons, but not neurogenesis, underlie the formation of granule cell dispersion in the epileptic hippocampus. J Neurosci 26:4701-4713.

Howell BW, Hawkes R, Soriano P, Cooper JA (1997) Neuronal position in the developing brain is regulated by mouse disabled-1. Nature 389:733-737.

Ikeda M, Mackay KB, Dewar D, McCulloch J. (1993) Differential alterations in adenosine A1 and kappa 1 opioid receptors in the striatum in Alzheimer's disease. Brain Res. 616:211-7.

Jaarsma D, Sebens JB, Korf J. (1991) Reduction of adenosine A1-receptors in the perforant pathway terminal zone in Alzheimer hippocampus. Neurosci Lett 121:111-114.

Jansen KL, Faull RL, Dragunow M, Synek BL. (1990) Alzheimer's disease: changes in hippocampal N-methyl-D-aspartate, quisqualate, neurotensin, adenosine, benzodiazepine, serotonin and opioid receptors--an autoradiographic study. Neuroscience 39:613-27.

Kalaria RN, Sromek S, Wilcox BJ, Unnerstall JR. (1990) Hippocampal adenosine A1 receptors are decreased in Alzheimer's disease. Neurosci Lett. 118:257-60.

Lambert de Rouvroit C, Goffinet AM (1998) The reeler mouse as a model of brain development. Adv Anat Embryol Cell Biol 150:1-106.

Lopez-Bendito G, Molnar Z (2003) Thalamocortical development: how are we going to get there? Nat Rev Neurosci 4:276-289.

Lujan R, Shigemoto R, Lopez-Bendito G (2005) Glutamate and GABA receptor signalling in the developing brain. Neuroscience 130:567-580.

Mariani J, Crepel F, Mikoshiba K, Changeux JP, Sotelo C. (1977) Anatomical, physiological and biochemical studies of the cerebellum from Reeler mutant mouse. Philos Trans R Soc Lond B Biol Sci. 281(978):1-28.

Marrone MC, Marinelli S, Biamonte F, Keller F, Sgobio CA, Ammassari-Teule M, Bernardi G, Mercuri NB (2006) Altered cortico-striatal synaptic plasticity and related behavioural impairments in reeler mice. Eur J Neurosci 24:2061-2070.

Matsuzaki H, Minabe Y, Nakamura K, Suzuki K, Iwata Y, Sekine Y, Tsuchiya KJ, Sugihara G, Suda S, Takei N, Nakahara D, Hashimoto K, Nairn AC, Mori N, Sato K. (2007) Disruption of reelin signaling attenuates methamphetamine-induced hyperlocomotion. Eur J Neurosci 25:3376-84.

Müller MC, Osswald M, Tinnes S, Haussler U, Jacobi A, Förster E, Frotscher M, Haas CA (2009) Exogenous reelin prevents granule cell dispersion in experimental epilepsy. Exp Neurol 216:390-397.

Niu S, Yabut O, D'Arcangelo G (2008) The Reelin signaling pathway promotes dendritic spine development in hippocampal neurons. J Neurosci 28:10339-10348.

Palomero-Gallagher N, Bidmon H-J, Cremer M, Schleicher A, Kircheis G, Reifenberger G, Kostopoulos G, Häussinger D, Zilles K (2009) Neurotransmitter receptor imbalances in motor cortex and basal ganglia in hepatic encephalopathy. Cell Physiol Biochem 24:291-306.

Paxinos G, Franklin KBJ (2001) The Mouse Brain in Stereotaxic Coordinates. Second Edition, Academic Press, San Diego, USA

Prestwich SA, Forda SR, Dolphin AC (1987) Adenosine antagonists increase spontaneous and evoked transmitter release from neuronal cells in culture. Brain Res 405:130-139.

Qiu S, Zhao LF, Korwek KM, Weeber EJ (2006) Differential reelin-induced enhancement of NMDA and AMPA receptor activity in the adult hippocampus. J Neurosci 26:12943-12955.

Rakic P, Caviness VS, Jr. (1995) Cortical development: view from neurological mutants two decades later. Neuron 14:1101-1104.

Saéz-Valero J, Fodero LR, Sjogren M, Andreasen N, Amici S, Gallai V, Vanderstichele H, Vanmechelen E, Parnetti L, Blennow K, Small DH (2003) Glycosylation of acetylcholinesterase and butyrylcholinesterase changes as a function of the duration of Alzheimer's disease. J Neurosci Res 72:520-526.

Sheldon M, Rice DS, D'Arcangelo G, Yoneshima H, Nakajima K, Mikoshiba K, Howell BW, Cooper JA, Goldowitz D, Curran T (1997) Scrambler and yotari disrupt the disabled gene and produce a reeler-like phenotype in mice. Nature 389:730-733.

Sheppard AM, Pearlman AL (1997) Abnormal reorganization of preplate neurons and their associated extracellular matrix: an early manifestation of altered neocortical development in the reeler mutant mouse. J Comp Neurol 378:173-179.

Simmons PA, Lemmon V, Pearlman AL (1982) Afferent and efferent connections of the striate and extrastriate visual cortex of the normal and reeler mouse. J Comp Neurol 211:295-308.

Simmons PA, Pearlman AL (1983) Receptive-field properties of transcallosal visual cortical neurons in the normal and reeler mouse. J Neurophysiol 50:838-848.

Sinagra M, Verrier D, Frankova D, Korwek KM, Blahos J, Weeber EJ, Manzoni OJ, Chavis P (2005) Reelin, very-low-density lipoprotein receptor, and apolipoprotein E receptor 2 control somatic NMDA receptor composition during hippocampal maturation in vitro. J Neurosci 25:6127-6136.

Stanfield BB, Cowan WM (1979) The morphology of the hippocampus and dentate gyrus in normal and reeler mice. J Comp Neurol 185:393-422.

Stone TW, Ceruti S, Abbracchio MP. (2009) Adenosine receptors and neurological disease: neuroprotection and neurodegeneration. Handb Exp Pharmacol 193:535-87.

Strazielle C, Hayzoun K, Derer M, Mariani J, Lalonde R (2006) Regional brain variations of cytochrome oxidase activity in Relnrl-orl mutant mice. J Neurosci Res 83:821-831.

Takahashi T, Goto T, Miyama S, Nowakowski RS, Caviness VS, Jr. (1999) Sequence of neuron origin and neocortical laminar fate: relation to cell cycle of origin in the developing murine cerebral wall. J Neurosci 19:10357-10371.

Terashima T, Inoue K, Inoue Y, Mikoshiba K (1987) Thalamic connectivity of the primary motor cortex of normal and reeler mutant mice. J Comp Neurol 257:405-421.

Trommsdorff M, Gotthardt M, Hiesberger T, Shelton J, Stockinger W, Nimpf J, Hammer RE, Richardson JA, Herz J (1999) Reeler/Disabled-like disruption of neuronal migration in knockout mice lacking the VLDL receptor and ApoE receptor 2. Cell 97:689-701.

Ułas J, Brunner LC, Nguyen L, Cotman CW Reduced density of adenosine A1 receptors and preserved coupling of adenosine A1 receptors to G proteins in Alzheimer hippocampus: a quantitative autoradiographic study.Neuroscience 52:843-54.

Ware ML, Fox JW, Gonzalez JL, Davis NM, Lambert de Rouvroit C, Russo CJ, Chua SC, Jr., Goffinet AM, Walsh CA (1997) Aberrant splicing of a mouse disabled homolog, mdab1, in the scrambler mouse. Neuron 19:239-249.

Weeber EJ, Beffert U, Jones C, Christian JM, Forster E, Sweatt JD, Herz J (2002) Reelin and ApoE receptors cooperate to enhance hippocampal synaptic plasticity and learning. J Biol Chem 277:39944-39952.

Wirths O, Multhaup G, Czech C, Blanchard V, Tremp G, Pradier L, Beyreuther K, Bayer TA. (2001) Reelin in plaques of beta-amyloid precursor protein and presenilin-1 double-transgenic mice. Neurosci Lett 316:145-148.

Zhao S, Chai X, Forster E, Frotscher M (2004) Reelin is a positional signal for the lamination of dentate granule cells. Development 131:5117-5125.

Zilles K, Amunts K (2009) Receptor mapping: architecture of the human cerebral cortex. Curr Opin Neurol 22:331-339.

Zilles K, Palomero-Gallagher N, Schleicher A (2004) Transmitter receptors and functional anatomy of the cerebral cortex. J Anat 205:417-432.

Zilles K, Qu MS, Kohling R, Speckmann EJ (1999) Ionotropic glutamate and GABA receptors in human epileptic neocortical tissue: quantitative in vitro receptor autoradiography. Neuroscience 94:1051-1061.

Zilles K, Schleicher A, Palomero-Gallagher N, Amunts K (2002) Quantitative analysis of cyto- and receptor architecture of the human brain. In: Brain Mapping. The Methods (Toga AW, Mazziotta JC, eds), pp 573-602. Amsterdam: Elsevier.

# Discussion

Changes of neurotransmitter receptor densities, distribution, and subunit expression were investigated in rodent models of neurodegeneration. Particularly, the effects of repeated seizures, inhibition of the glutamate/glutamine cycle or reelin mutation were analyzed. To allow a more differential study of neurotransmitter receptor subunits on the mRNA level a new method of in situ hybridization was established.

## 1. Pentylenetetrazole-induced seizures

Repetitive intraperitoneal injection of PTZ in rats induced acute and chronic seizures, resulting in an oxidative stress response as indicated by enhanced heat shock protein 27 (HSP-27) expression in astrocytes (Bidmon et al. 2008). Since astrocytes decisively contribute to neurotransmitter metabolism, we hypothesized imbalances of neurotransmitter receptor expression in this seizure model. PTZ induced seizures resulted in a general decrease of kainate receptor densities, together with an increase of NMDA binding sites in the hippocampus, the somatosensory, piriform and the entorhinal cortices. Furthermore, $A_1$ binding sites were decreased in the amygdala and CA1, while BZ binding sites were increased in the dentate gyrus and CA1. However, the observed changes were independent of the glial stress response, exhibiting a much wider regional distribution than the enhanced glial HSP-27 expression.

Binding site densities for kainate were significantly reduced in all investigated regions of PTZ treated rats. Systemic administration of kainate has been extensively used as a model for temporal lobe epilepsy (reviewed by Vincent and Mulle 2009). However, due to the varying synaptic and subcellular localization and functions of kainate binding sites (reviewed by Lerma, 2003; Pinheiro and Mulle, 2006), it is still a matter of debate how this receptor class is involved in the epileptic pathology. Here, high affinity kainate receptors (Monaghan and Cotman, 1982) with binding sites located at the KA1- and KA2-subunits (Werner, 1991; Herb, 1992) were quantified, which reversibly inhibit $K^+$-currents responsible for the slow after-hyperpolarizing potential in hippocampal pyramidal cells (Gho et al., 1986; Melyan et al., 2002; Fisahn et al., 2005; Ruiz et al., 2005). Thus, an activation of kainate receptors results in enhanced neuronal excitability at least in the hippocampal formation. Since we demonstrated a reduction of kainate binding in all investigated brain regions of PTZ-treated rats, one could speculate that decreased kainate binding acts as mechanism to reduce enhanced neuronal function in the epileptic brain. However, further effort is necessary to elucidate this subject. Particularly, it has to be investigated whether the observed changes are due to PTZ-treatment (i.e. $GABA_A$ inhibition) or the reduction of kainate binding site is secondary to seizure onset.

Though PTZ is a potent inhibitor of $GABA_A$ function, densities of the $GABA_A$ receptor were unaffected in our model. However, we did find an increase of BZ binding site densities in the CA1 region and the dentate gyrus of PTZ-treated rats. The BZ binding site is located between $\gamma_2$ and $\alpha_{(1,2,3,5)}$ subunits of the pentameric $GABA_A$ receptor, while GABA (as well as the [$^3$H]-ligand muscimol) binds between $\alpha$ and $\beta$ subunits of the receptor (Pritchett et al., 1989; Rudolph et al., 1999; McKernan et al., 2000; Baumann et al., 2003). Thus, these results indicate alterations of the subunit-composition of the functional $GABA_A$ receptor at least in these areas. As the hippocampus is known to be especially susceptible to epileptic activity (Majores et al., 2007; Scharfman and Gray, 2007) and it has further been shown that $GABA_A$ agonist administration reduces the effect of PTZ-application after single as well as chronic treatment (Hansen et al., 2004), this upregulation might generally attenuate neuronal activity in this regions.

HSP-27 expression is regionally elevated in astrocytes due to oxidative stress in both human epileptic patients as well as in the PTZ-model (Bidmon et al., 2004, 2008). Therefore, one goal of the present study was to test the hypothesis that PTZ-induced impairment of glia cells results in alterations of neurotransmitter receptor densities. We found a partial overlap of regions showing glial heat shock response and altered receptor densities. Regions in which altered HSP-27 expression and binding site densities coincided were the piriform-entorhinal cortex, the dentate gyrus and the amygdala. In the somatosensory cortex receptor binding sites were significantly changed, but not the glial heat shock response (Bidmon et al., 2005). For the retrosplenial cortex we found a decrease of kainate binding but no increase in HSP-27 induction (Bidmon et al. 2005). Therefore, changes in the retrosplenial cortex would correspond to the early activations observed during PTZ-induced seizures as revealed by functional imaging (Brevard et al. 2006) or c-fos induction (André et al., 1998). However, the most striking difference was detected in CA1, where considerable alterations of receptor binding sites were not accompanied by a glial heat shock response. We therefore suggest, that alterations of receptor densities in the PTZ-model occur independent of corresponding glial impairment and precede the glial reaction in a region-specific manner.

Rats were treated with the convulsive PTZ to investigate the effects of repeated seizures on neurotransmitter receptor densities. Changes of binding site densities for kainate, NMDA, $A_1$ and BZs were demonstrated, which were independent of a hypothesized concomitant HSP-27 expression. Thus, neurotransmitter receptor alterations in the PTZ model did not follow a glial stress response.

## 2. Inhibition of glutamine synthetase

The in vivo inhibition of GS in rats resulted in decreased BZ binding site densities in the somatosensory, piriform, and entorhinal cortices and in the amygdala 24h and 72h after MSO-treatment. Additionally, decreased BZ-binding on cerebral membrane homogenates was demonstrated 72h after MSO-treatment without changes in binding site affinity. Furthermore, we found differential changes of $GABA_A$ $\alpha_1$, $\gamma_2$ and phosphorylated $\gamma_2$ (p- $\gamma_2$) subunits following MSO treatment. These changes were independent from alterations of the respective mRNA levels as revealed by quantitative in situ hybridization. Inhibition of GS did not affect glial glutamate transporter GLAST and Glt-1 quantities.

In MSO treated rats changes of BZ binding site densities and GABAergic subunit composition comprise a regional and temporal dependency. Thus, inhibition of the GGC does not result in a distinct alteration of GABAergic receptors, but is moreover regulated by spatial and temporal factors, the most likely of which is seizure susceptibility (within the epileptic circuitry). Results from autoradiography experiments exhibited decreased densities of BZ binding sites 24h and 72h after MSO treatment. This decrease only proved to be significant in the dentate gyrus 24h after treatment, and after 24h as well as 72h in the amygdala, the piriform and entorhinal cortices. Interestingly, the piriform and entorhinal cortices are key regions within the epileptic circuitry (White, 2002). Since MSO affects all astrocytes (Bidmon et al., 2008), this regional specificity indicates that MSO-induced seizures differentially affect the neuron-glia interaction in seizure prone cerebral regions. Interestingly, this pattern basically corresponds to an enhanced expression of HSP 27 (a label for affected astrocytes) in MSO treated rats and, further, to enhanced GS-nitration following pentylenetetrazole (GABA-antagonist)-induced seizures (Bidmon et al., 2008). Therefore, we propose that the piriform/entorhinal cortices, dentate gyrus and hippocampal CA1 region are key regions in which impaired astrocytic glutamate metabolism causes region-specific changes of neurotransmitter receptors in seizure prone neurons and/or glia cells in both animal models.

During the course of GS-inhibition we found decreased levels of $α_1$ subunits after 24h but not after 72h. Furthermore, we measured a decrease of $γ_2$ and phosphorylated $γ_2$ (p-$γ_2$) subunits after 24h, but an increase of $γ_2$ without changes of p-$γ_2$ after 72h. Conversely, quantities of functional BZ binding sites in brain homogenates were unchanged after 24h but significantly decreased after 72h. Thus, it seems most likely that subunit expression alterations precede functional binding site alterations at the multimeric receptor. $GABA_A$ receptor subunit expression and ligand binding have previously been demonstrated to be differentially altered in rodent models of temporal lobe epilepsy (e.g. Walsh et al., 1999; Volk et al., 2006; Bethmann et al., 2008). For example, the expression of $α_1$, $α_2$, $β_{2/3}$ and $γ_2$ subunits was significantly decreased in the hippocampus of antiepileptic drug resistant rats (Bethmann et al., 2008). Therefore, it is tempting to speculate that $GABA_A$ subunit changes induced by GGC disruption might be key events for both epileptogenesis (Eid et al., 2004; 2008) as well as antiepileptic drug resistance (reviewed by Löscher 2009; Schmidt and Löscher 2009).

Alterations of glutamate recycling could result in changes of astrocytic glutamate transport (Rothstein and Tabakoff, 1985). Therefore, we investigated the expression of the major glutamate transporters GLAST and Glt-1 in MSO treated rats versus controls. The astrocytic membrane is responsible for at least 80% of glutamate clearance and the majority of synaptic inactivation in the brain (Bergles and Jahr, 1997; Danbolt, 2001). An increase of glutamate concentration in the synaptic cleft could induce excitotoxicity. However, we did not observe evidence for enhanced cellular decline in our model. This observation is supported by experiments on the effect of MSO (350 mg/kg) on blood brain barrier permeability, where ultrastructural investigation demonstrated that MSO, even in high concentrations, does not affect glial integrity (Nitsch et al., 1986) and pathological changes in glia cells are induced only after repeated application of MSO. Astrocytic glutamate uptake is significantly enhanced for up to 7 days after intraventricular injection of MSO (Rothstein and Tabakoff, 1984). Therefore, we hypothesized that an enhanced glial glutamate transport might lead to an increased transporter expression. GLAST and Glt-1 are the predominant glutamate transporters in the brain and both are mainly expressed by astrocytes (Gadea and López-Colomé, 2001). However, we did not find changes of either GLAST or Glt-1 expression in our study. Thus, available glutamate transporters sufficiently remove synaptic glutamate, or an enhanced transport is not mediated by increased transporter expression, but rather by changes in removal rate or affinity (Rothstein and Tabakoff, 1984).

The astrocytic enzyme glutamine synthetase is a key regulator of glutamate and GABA metabolism. We demonstrated that in vivo inhibition of GS induces differential changes of BZ binding and $GABA_A$ receptor subunit composition. In conclusion, our results suggest a feedback regulation between neurotransmitter recycling in the GGC and GABAergic synaptic transmission, likely resulting in altered $GABA_A$ modulation by BZs.

## 3. Reeler mice

Neurotransmitter receptor densities and distribution were significantly altered in reeler mouse brains when compared to the wild type. In the cerebral cortex some receptor subtypes demonstrated an obvious laminar inversion (i.e. AMPA, kainate, NMDA, $GABA_B$, $5-HT_1$, $M_1$, $M_3$, nAch), while other subtypes ($A_1$, $GABA_A$, BZ, $5-HT_2$, $M_2$, $\alpha_1$, $\alpha_2$) were less strikingly affected. A significant decrease of $B_{max}$ in whole brain homogenates was found for $A_1$ and $GABA_A$ receptors. The later was demonstrated to be restricted to the cerebellum, while $A_1$ receptors were also significantly reduced in all investigated regions of the forebrain (except CA1). Several binding site densities exhibited regionally restricted changes in the forebrain (i.e. kainate, $A_1$, benzodiazepine, $5-HT_1$, muscarinic $M_2$, adrenergic $\alpha_1$ and $\alpha_2$), while other receptor types were unaffected in the investigated regions (i.e. AMPA, NMDA, $GABA_A$, $5-HT_2$, $D_1$, M1, M3, nAch).

In reeler mice, a disturbed developmental migration of neurons has been shown to result in an inversion of neocortical layers (Caviness, 1982; reviewed by Frotscher, 1998; Rice and Curran, 2001). However, the assumption of a "simple" inversion of neocortical layers in the reeler brain is still controversially discussed. For example, injection of horseradish peroxidase (HRP) into the thalamus labeled thalamocortical neurons in the layer VI of wild type mice, whereas the same experiment in reeler mice resulted in about 60% of retrogradely labeled neurons localized beneath the pial surface and the remaining labeled neurons evenly scattered along the radial axis of the cortex (Yamamoto et al., 2003). On the other hand, it was suggested that thalamocortical input into the barrel field of the somatosensory cortex is generally maintained in reeler, although barrels were shown to be structurally altered (reviewed by Lopez-Bendito and Molnar, 2003). Earlier studies demonstrated that when HRP was injected into the spinal cord to retrogradely label the corticospinal tract, the cell soma of the pyramidal neurons in the layer V of the cerebral cortex were stained in the wild type, while in reeler, the retrogradely labeled corticospinal neurons distributed widely throughout the radial cortical depth (Terashima et al., 1983). Thus, it can not be concluded doubtlessly that the spatial relationship at least of layer V and layer VI neurons is inverted in the reeler cortex. In our study of neurotransmitter receptor distribution in the somatosensory cortex of wild type and reeler brains, we found that AMPA, kainate, NMDA, $GABA_B$, $M_1$, $M_3$, nAch and $5-HT_1$ receptor densities reflected an inverted distribution in reeler, while other receptor subtypes ($GABA_A$, BZ, $M_2$, $5-HT_{2A}$, $\alpha_1$, $\alpha_2$ and $A_1$) did not show a clearly inverted distribution pattern. Therefore, we suggest that the reeler cortex may basically be inverted, however, not all receptor types were affected to the same degree. Therefore, our results support both hypotheses of an inverted vs. a scattered neocortical organization, depending on the receptor type emphasized. These observations are partially confirmed by Dekimoto et al. (2010) who used *in situ* hybridization to study the distribution of layer specific mRNAs in *reeler* and control brains. Similar to our findings, they reported that layer specificity in the *reeler* somatosensory cortex exhibits a complex pattern ranging from splitting of layers (layer VI, characterized by Tbr1 expression) to a complete disruption of layer integrity and dispersion of the respective neurons (layer V, characterized by ER81 expression).

Here, we demonstrated differential changes of neurotransmitter receptors in all investigated regions of the reeler brain. While most receptor changes were locally restricted, the density of $A_1$ receptors was decreased in all regions of the forebrain (except CA1). Additionally, this decrease of $A_1$ density was verified on crude neuronal membrane preparations of reeler whole brain homogenates.

The function of reelin in the adult brain has been demonstrated to be involved in synaptic plasticity by modulation of glutamate receptor function (reviewed by Rogers and Weeber, 2008). For example, reelin perfusion was shown to enhance synaptic plasticity in hippocampal slices after high frequency stimulation of Shaffer collaterals (Weeber et al., 2002). This enhancement of long term potentiation is mediated by phosphorylation of NR2A subunits of the NMDA receptor, thereby increasing $Ca^{2+}$

Discussion

conductance (Beffert et al., 2005). Furthermore, longer reelin perfusion (> 20 minutes) resulted in an NMDA receptor independent increase of synaptic AMPA receptors in CA1 (Qiu et al., 2006).

Due to the influence of reelin on NMDA and AMPA receptors in vitro, changes of these receptors could have been expected in reeler mice. Although we did not observe alterations of AMPA or NMDA receptor density or affinity, we found a massive decrease of $A_1$ receptor densities in reeler. Activation of $A_1$ receptors results in a reduction of synaptic glutamate release (Prestwich et al., 1987), thereby regulating neuronal activity, synaptic plasticity and learning (reviewed by Fredholm et al., 2005; Stone et al., 2009). Thus, it can be hypothesized that a lack of reelin and, therefore, a lack of its enhancing effect on glutamatergic AMPA and NMDA receptors, might be partially compensated by a downregulation of $A_1$ receptors. A decreased activation of $A_1$ receptors would result in an enhanced glutamate release (Prestwich et al., 1987). It would be an interesting approach to investigate the effect of reelin enriched medium on $A_1$ receptor expression in organotypic slice culture of wild type an reeler brains. An inhibition of reelin function in the wild type using the CR-50 antibody (Ogawa et al., 1995; D'Arcangelo et al., 1997; Miyata et al., 1997; Nakayima et al., 1997) would complete these experiments. Additionally, a comprehensive receptor autoradiographic study in heterozygous reeler mice, which express approximately 50% less reelin than the wild type (Tueting et al., 1999; Liu et al., 2001), could yield evidence on the in vivo effect of reduced reelin expression on neurotransmitter receptors. Furthermore, this would rule out the possibility, that the observed changes of neurotransmitter receptors in reeler are secondary to the disturbed neuronal organization rather than to a lack of the functional protein. However, these questions remain subject of further investigation.

The extracellular matrix protein reelin regulates neuronal development and modulates neurotransmitter receptor activity in the adult. Here, we demonstrated differential changes of neurotransmitter receptor densities and distribution in reeler mice, lacking the functional protein. Our results partially supplement on the assumption of a neocortical inversion in reeler and support previous hypotheses of a role for reelin in neurotransmitter receptor regulation.

Discussion

## 4. Quantitative in situ hybridization

A new method of qISH was established using $^{33}$P-labelled deoxyribonucleotides in combination with $^{14}$C polymer standards and a phosphorus imaging system. To test our approach we quantified the mRNAs of the glutamatergic AMPA receptor subunits GluR1 and GluR2 in the hippocampus of untreated rats, and after application of DMA$^{III}$. According to the hypothesis, the mRNA levels of GluR1 and GluR2 were significantly reduced in DMA$^{III}$ treated rats.

Since conventional film autoradiography requires exposure times ranging from several days to months, we were interested in reducing the time required for this procedure by utilizing phosphorus screen imaging. Moreover, dose response of an autoradiography film is linear only over a rather narrow range of exposure, which further limits the utility of film autoradiography. The use of phosphorus imaging is a suitable approach to generate autoradiographs in qISH within only a few hours or days. Depending on specific activity of the hybridization probe, images of high quality can be produced in less than 8 h. However, phosphorus imaging plates exhibit a technically limited resolution of only 50 µm, thus, resulting in a apparent disadvantage compared to conventional film autoradiography. A resolution of approximately 10 µm can be achieved using conventional film and $^{3}$H-labelled probes in classic receptor autoradiography (Zilles et al., 2002). However, the use of $^{3}$H-labelling is generally regarded as inappropriate for qISH, since quenching of $^{3}$H radiation by lipophilic structures and proteins within the tissue would adulterate quantitative analysis. Thus, probes for qISH are typically labeled using $^{35}$S or $^{33}$P, which are less affected by tissue-inherent quenching due to their higher emission energies. Our results demonstrated a maximum spatial resolution between 140-160 µm within a $^{33}$P-labelled tissue sample. Conclusively, when working with $^{33}$P, there is no disadvantage in image resolution using phosphorus imaging plates.

We further simplified the experimental workflow of qISH. To avoid the time consuming preparation of $^{33}$P tissue standards for each experiment, which is needed for a reliable densitometric analysis, we used commercially available $^{14}$C-standards. Polymer standards labeled with $^{14}$C can be used for calibration of autoradiographs prepared with $^{35}$S or $^{33}$P-labelled probes (Miller, 1991; Baskin and Stahl, 1993), since these isotopes exhibit ß$^-$-radiation with similar emission spectra. The use of $^{14}$C is beneficial because of its half-life (5730 years) compared to short-lived $^{33}$P (25.4

days). Thus, $^{14}$C-polymer standards can be repeatedly used in hybridization experiments using $^{33}$P-labelled probes.

Our results showed that single treatment of rats with DMA$^{III}$ significantly reduces the mRNA expression levels of the AMPA receptor subunits GluR1 and GluR2 in the hippocampus. Since DMA$^{III}$ treatment was previously demonstrated to reduce the number of hippocampal AMPA receptors (Lopez Escobar, 2009; Krüger et al., 2006; 2007) the mRNA of subunits GluR1 and GluR2 showed the awaited reduction after using the present method. Thus, our method describes a reliable technique that can be easily used in any laboratory to quantify the expression of high and low abundant mRNAs.

## 5. Summary and conclusions

In the present work, neurotransmitter receptors were studied in rodent models of repeated seizures, inhibition of the glutamate/glutamine cycle or reelin gene mutation, respectively. While the specific results were discussed above with emphasis on the underlying hypotheses of each study, in the following these findings will be briefly reflected according to common aims and questions underlying all studies.

1. Are the densities and regional distribution of neurotransmitter receptors affected in a given animal model?

Significant changes of neurotransmitter receptor densities were demonstrated in all investigated models, most of which exhibited a complex temporal and spatial pattern of these alterations. Several receptor types were differentially changed in the PTZ model as well as in the reeler mouse brain, while inhibition of GS in rats exclusively resulted in decreased BZ binding site densities. A disturbed cytoarchitecture in the reeler mouse brain was accompanied by changes of laminar receptor distribution, complementing studies of the reeler phenotype and partially supporting the hypothesis of an inverted neocortical organization in reeler. Thus, these results demonstrate that neurotransmitter receptor changes correlate with disturbances of functional as well as structural determinants of normal brain function in different animal models.

Discussion

2.  Are the observed changes restricted to a single receptor type or a specific brain region?

In the present study, alterations of a specific receptor type were most often accompanied by changes of other receptors systems. Correlating changes could occur in several brain regions, or be otherwise regionally restricted. For example, in the PTZ model a general decrease of kainate binding site densities was accompanied by increased NMDA receptor densities in most (i.e. five out of seven) but not all investigated regions. On the other hand, a global decrease of $A_1$ receptors in reeler mice was accompanied by decreased adrenergic $\alpha_1$ and $\alpha_2$ densities in the motor cortex only. The inhibition of GS in rats exclusively resulted in decreased BZ binding sites, however this decrease was demonstrated in several brain regions. Together, these observations underline the necessity of multiple receptor studies for investigations on receptor expression under different experimental conditions.

3.  Do changes of receptor density correlate with altered receptor pharmacology (e.g. binding site affinity, maximum binding capacity)?

The maximum binding capacity and binding affinity of neurotransmitter receptors were investigated using saturation analysis of brain homogenates in reeler mice and MSO treated rats. In both models significant changes of $B_{max}$ were found for some receptors, while the respective $K_D$ values were unaffected in all cases. The observed changes of $B_{max}$ could be verified by respective regional density changes of the receptors in native brain sections. Thus, in this study the affinity of the investigated receptors were maintained in reeler mice and after GS inhibition in rats, while $B_{max}$ values confirmed changes of regional receptor densities as revealed by receptor autoradiography of native brain sections.

4.  Do changes of receptor subunit composition or mRNA transcription levels play a role in receptor regulation?

A new method of quantitative in situ hybridization was established. Using this approach it was demonstrated that decreased AMPA receptor densities in the hippocampus of DMA[III] treated rats correlated with decreased receptor subunit mRNA expression. Inhibition of GS resulted in differential changes of $GABA_A$ receptor subunit composition as revealed by Western blot analysis. However, these changes were independent of a respective decrease of subunit mRNAs. Thus, changes of receptor subunit composition and expression were shown, which did not depend inevitably on changes of the respective mRNA levels.

Taken together, the investigation of neurotransmitter receptors in rodent models of repeated seizures, inhibition of the glutamate/glutamine cycle or reelin gene mutation, respectively, demonstrated that i) loss of neuronal structure or function coincided with differential changes of neurotransmitter receptor densities in all investigated models ii) correlating changes of receptor densities could occur in numerous brain regions or in a regionally restricted manner iii) a disturbed neuronal function could influence receptor subunit composition and mRNA expression.

Conclusively, this study revealed a complex pattern of correlations between disturbances of neuronal structure or function and changes of neurotransmitter receptors. Since neurotransmitter receptors are a major target for pharmacological intervention, these results might offer ambitions for the development of new therapeutic strategies.

## References

André V, Pineau N, Motte JE, Marescaux C, Nehlig A (1998) Mapping of neuronal networks underlying generalized seizures induced by increasing doses of pentylenetetrazol in the immature and adult rat: a c-Fos immunohistochemical study. Eur J Neurosci. 10:2094-2106.

Angelatou F, Pagonopoulou O, Kostopoulos G (1990) Alterations of $A_1$ adenosine receptors in different mouse brain areas after pentylentetrazol-induced seizures, but not in the epileptic mutant mouse 'tottering'. Brain Res 534:251-256.

Baskin DG, Stahl WL (1993) Fundamentals of quantitative autoradiography by computer densitometry for in situ hybridization, with emphasis on 33P. J Histochem Cytochem 41:1767-1776.

Baumann SW, Baur R, Sigel E (2003) Individual properties of the two functional agonist sites in $GABA_A$ receptors. J Neurosci 23:11158-11166.

Beffert U, Weeber EJ, Durudas A, Qiu S, Masiulis I, Sweatt JD, Li WP, Adelmann G, Frotscher M, Hammer RE, Herz J. (2005) Modulation of synaptic plasticity and memory by Reelin involves differential splicing of the lipoprotein receptor Apoer2. Neuron 47:567-79.

Belelli D, Lambert JJ (2005) Neurosteroids: endogenous regulator of the GABA(A) receptor. Nat Rev Neurosci 6:565-575.

Bergles DE, Jahr CE (1997) Synaptic activation of glutamate transporters in hippocampal astrocytes. Neuron 19:1297-1308.

Bertram E (2007) The relevance of kindling for human epilepsy. Epilepsia 48 Suppl 2:65-74.

Bethmann K, Fritschy JM, Brandt C, Löscher W (2008) Antiepileptic drug resistant rats differ from drug responsive rats in GABA(A) receptor subunit expression in a model of temporal lobe epilepsy. Neurobiol Dis. 31:169-87.

References

Bidmon HJ, Görg B, Palomero-Gallagher N, Behne F, Lahl R, Pannek HW, Speckmann EJ, Zilles K (2004) Heat shock protein-27 is upregulated in the temporal cortex of patients with epilepsy. Epilepsia 45:1549-1559.

Bidmon HJ, Görg B, Palomero-Gallagher N, Schleicher A, Häussinger D, Speckmann EJ, Zilles K (2008) Glutamine synthetase becomes nitrated and its activity is reduced during repetitive seizure activity in the pentylentetrazole model of epilepsy. Epilepsia 49(10):1733-1748

Bidmon HJ, Gorg B, Palomero-Gallagher N, Schliess F, Gorji A, Speckmann EJ, Zilles K (2005) Bilateral, vascular and perivascular glial upregulation of heat shock protein-27 after repeated epileptic seizures. J Chem Neuroanat 30:1-16.

Bidmon HJ, Palomero-Gallagher N, Zilles K (2002) Postoperative Untersuchungen in der Epilepsiechirurgie: Enzyme der oxidativen Stresskaskade, Multi-Drug-Transporter und Transmitterrezeptoren. Klin Neurophysiol 33:168-177.

Brevard ME, Kulkarni P, King JA, Ferris CF (2006) Imaging the neural substrates involved in the genesis of pentylenetetrazol-induced seizures. Epilepsia 47:745-754.

Caspers H, Speckmann EJ (1972) Cerebral $pO_2$, $pCO_2$ and pH: changes during convulsive activity and their significance for spontaneous arrest of seizures. Epilepsia 13:699-725.

Cavines VS Jr (1982) Neocortical histogenesis in normal and reeler mice: a developmental study based on [3H]thymidine autoradiography. Brain Res 256:293-302

Chaudhry FA, Reimer RJ, Edwards RH (2002) The glutamine commute: take the N line and transfer to the A. J Cell Biol 157:349-355.

Cremer CM (2007) Receptor Archtitecture of the Reeler Mouse Brain. Diploma thesis, Institute for Neuroscience and Medicine (INM-2), Research Center Jülich.

Cremer CM, Bidmon HJ, Görg B, Palomero-Gallagher N, Escobar JL, Speckmann EJ, Zilles K (2010a) Inhibition of glutamate/glutamine cycle in vivo results in decreased benzodiazepine binding and differentially regulated GABAergic subunit expression in the rat brain. Epilepsia, in press

Cremer CM, Cremer M, Lopez Escobar J, Speckmann EJ, Zilles K. (2009b) Fast, quantitative in situ hybridization of rare mRNAs using (14)C-standards and phosphorus imaging. J Neurosci Methods 185:56-61.

Cremer CM, Lübke JH, Palomero-Gallagher N, Zilles K (2010b) Laminar distribution of neurotransmitter receptors in the *reeler* mouse cerebral cortex. J Neurosci, in review

Cremer CM, Palomero-Gallagher N, Bidmon HJ, Schleicher A, Speckmann EJ, Zilles K (2009a) Pentylenetetrazole-induced seizures affect binding site densities for GABA, glutamate and adenosine receptors in the rat brain. Neuroscience 163:490-499.

Danbolt NC (2001) Glutamate uptake. Prog Neurobiol 65:1-105

D'Arcangelo G, Nakajima K, Miyata T, Ogawa M, Mikoshiba K, Curran T (1997) Reelin is a secreted glycoprotein recognized by the CR-50 monoclonal antibody. J Neurosci 17:23-31.

Dekimoto H, Terashima T, Katsuyama Y (2010) Dispersion of the neurons expressing layer specific markers in the reeler brain. Develop Growth Differ 52:181-193.

Durakoglugil MS, Chen Y, White CL, Kavalali ET, Herz J (2009) Reelin signaling antagonizes beta-amyloid at the synapse. Proc Natl Acad Sci U S A 106:15938-15943.

Eid T, Ghosh A, Wang Y, Beckström H, Zaveri HP, Lee TS, Lai JC, Malthankar-Phatak GH, de Lanerolle NC (2008) Recurrent seizures and brain pathology after inhibition of glutamine synthetase in the hippocampus in rats. Brain 131:2061-2070.

Eid T, Thomas MJ, Spencer DD, Rundén-Pran E, Lai JC, Malthankar GV, Kim JH, Danbolt NC, Ottersen OP, de Lanerolle NC (2004) Loss of glutamine synthetase in the human epileptogenic hippocampus: possible mechanism for raised extracellular glutamate in mesial temporal lobe epilepsy. Lancet 363:28-37.

Ekonomou A, Angelatou F (1999) Upregulation of NMDA receptors in hippocampus and cortex in the pentylenetetrazol-induced "kindling" model of epilepsy. Neurochem Res 24:1515-1522.

Ekonomou A, Smith AL, Angelatou F (2001) Changes in AMPA receptor binding and subunit messenger RNA expression in hippocampus and cortex in the pentylenetetrazole-induced 'kindling' model of epilepsy. Brain Res Mol Brain Res 95:27-35.

Fatemi SH (2008) Reelin Glycoprotein: Structure, Biology and Roles in Health and Disease. Springer Science and Business Media, New York

Fisahn A, Heinemann SF, McBain CJ (2005) The kainate receptor subunit GluR6 mediates metabotropic regulation of the slow and medium AHP currents in mouse hippocampal neurones. J Physiol. 562:199-203.

Fonnum F, Paulsen RE (1990) Comparison of transmitter amino acid levels in rat globus pallidus and neostriatum during hypoglycemia or after treatment with methionine sulfoximine or gamma-vinyl gamma-aminobutyric acid. J Neurochem 54:1253-1257.

Fredholm BB, Chen JF, Masino SA, Vaugeois JM (2005b) Actions of adenosine at its receptors in the CNS: insights from knockouts and drugs. Annu Rev Pharmacol Toxicol 45:385-412.

Frotscher M (1998) Cajal-Retzius cells, Reelin, and the formation of layers. Curr Opin Neurobiol 8:570-575.

Gadea A, López-Colomé AM (2001) Glial transporters for glutamate, glycine and GABA I. Glutamate transporters. J Neurosci Res 63:453-60.

Gho M, King AE, Ben-Ari Y, Cherubini E (1986) Kainate reduces two voltage-dependent potassium conductances in rat hippocampal neurons in vitro. Brain Res. 385:411-414.

References

Hansen SL, Sperling BB, Sanchez C (2004) Anticonvulsant and antiepileptogenic effects of GABA$_A$ receptor ligands in pentylenetetrazole-kindled mice. Prog Neuropsychopharmacol Biol Psychiatry 28:105-113.

Herb A, Burnashev N, Werner P, Sakmann B, Wisden W, Seeburg PH (1992) The KA-2 subunit of excitatory amino acid receptors shows widespread expression in brain and forms ion channels with distantly related subunits. Neuron 8:775-785.

Herz J, Chen Y (2006) Reelin, lipoprotein receptors and synaptic plasticity. Nat Rev Neurosci 7:850-859.

Huang RQ, Bell-Horner CL, Dibas MI, Covey DF, Drewe JA, Dillon GH (2001) Pentylenetetrazole-induced inhibition of recombinant gamma-aminobutyric acid type A (GABA$_A$) receptors: mechanism and site of action. J Pharmacol Exp Ther 298:986-995.

Kandel ER, James H, Jessell M (2000) Principles of Neural Science, 4th edition, McGraw-Hill, New York.

Kocherhans S, Madhusudan A, Doehner J, Breu KS, Nitsch RM, Fritschy JM, Knuesel I (2010) Reduced Reelin expression accelerates amyloid-beta plaque formation and tau pathology in transgenic Alzheimer's disease mice. J Neurosci. 30:9228-9240.

Krüger K, Gruner J, Madeja M, Hartmann LM, Hirner AV, Binding N, Mushoff U (2006) Blockade and enhancement of glutamate receptor responses in Xenopus oocytes by methylated arsenicals. Arch Toxicol 80:492-501.

Krüger K, Repges H, Hippler J, Hartmann LM, Hirner AV, Straub H, Binding N, Mushoff U (2007) Effects of dimethylarsinic and dimethylarsinous acid on evoked synaptic potentials in hippocampal slices of young and adult rats. Toxicology and Applied Pharmacology 225:40-46.

Kvamme E, Roberg B, Torgner IA (2000) Phosphate-activated glutaminase and mitochondrial glutamine transport in the brain. Neurochem Res 25:1407-1419.

## References

Lamar C Jr, Sellinger OZ (1965) The inhibition in vivo of cerebral glutamine synthetase and glutamine transferase by the convulsant methionine sulfoximine. Biochem Pharmacol 14:489-506

Lerma J (2003) Roles and rules of kainate receptors in synaptic transmission. Nat Rev Neurosci. 4:481-495.

Levenson JM, Qiu S, Weeber EJ (2008) The role of reelin in adult synaptic function and the genetic and epigenetic regulation of the reelin gene. Biochim Biophys Acta 1779:422-431.

Liang SL, Carlson GC, Coulter DA (2006) Dynamic regulation of synaptic GABA release by the glutamate-glutamine cycle in hippocampal area CA1. J Neurosci. 26:8537-8548.

Liu WS, Pesold C, Rodriguez MA, Carboni G, Auta J, Lacor P, Larson J, Condie BG, Guidotti A, Costa E (2001) Down-regulation of dendritic spine and glutamic acid decarboxylase 67 expressions in the reelin haploinsufficient heterozygous reeler mouse. Proc Natl Acad Sci U S A 98:3477-3482.

Löscher W (2009) Molecular mechanisms of drug resistance in status epilepticus. Epilepsia 50 Suppl 12:19-21.

Lopez-Bendito G, Molnar Z (2003) Thalamocortical development: how are we going to get there? Nat Rev Neurosci 4:276-289.

Lopez Escobar J, 2009, Influence of treatment with arsen- and tinorganic compounds in early ontogenesis (P6) on the functional expression of glutamatergic and GABAergic receptors in the hippocampus of the rat. Diploma thesis, Institute for Neuroscience and Medicine (INM-2), Research Center Jülich.

Macdonald RL, Barker JL (1978) Specific antagonism of GABA-mediated postsynaptic inhibition in cultured mammalian spinal cord neurons: a common mode of convulsant action. Neurology 28:325-330.

Majores M, Schoch S, Lie A, Becker AJ (2007) Molecular neuropathology of temporal lobe epilepsy: complementary approaches in animal models and human disease tissue. Epilepsia 48 Suppl 2:4-12.

# References

Martin DL, Tobin AJ (2000) Mechanisms controlling GABA synthesis and degradation in the brain. In: GABA in the nervous system (Martin DL, Olsen RW, eds), pp 25-41. Philadelphia: Lippincott Williams & Wilkins.

Martinez-Hernandez A, Bell KP, Norenberg MD (1977) Glutamine synthetase: glial localization in brain. Science 195:1356-1358.

Miller JA (1991) The calibration of 35S or 32P with 14C-labeled brain paste or 14C-plastic standards for quantitative autoradiography using LKB Ultrofilm or Amersham Hyperfilm. Neurosci Lett 121:211-214.

Miyata T, Nakajima K, Mikoshiba K, Ogawa M (1997) Regulation of Purkinje cell alignment by reelin as revealed with CR-50 antibody. J Neurosci 17:3599-609.

Nakajima K, Mikoshiba K, Miyata T, Kudo C, Ogawa M (1997) Disruption of hippocampal development in vivo by CR-50 mAb against reelin. Proc Natl Acad Sci U S A 94:8196-8201.

Nitsch C, Goping G, Klatzo I (1986) Pathophysiological aspects of blood-brain barrier permeability in epileptic seizures. Adv Exp Med Biol. 203:175-189.

McKernan RM, Rosahl TW, Reynolds DS, Sur C, Wafford KA, Atack JR, Farrar S, Myers J, Cook G, Ferris P, Garrett L, Bristow L, Marshall G, Macaulay A, Brown N, Howell O, Moore KW, Carling RW, Street LJ, Castro JL, Ragan CI, Dawson GR, Whiting PJ (2000) Sedative but not anxiolytic properties of benzodiazepines are mediated by the $GABA_A$ receptor alpha1 subtype. Nat Neurosci 3:587-592.

Melyan Z, Wheal HV, Lancaster B (2002) Metabotropic-mediated kainate receptor regulation of IsAHP and excitability in pyramidal cells. Neuron 34:107-114.

Monaghan DT, Cotman CW (1982) The distribution of [3H]kainic acid binding sites in rat CNS as determined by autoradiography. Brain Res. 252:91-100.

Ogawa M, Miyata T, Nakajima K, Yagyu K, Seike M, Ikenaka K, Yamamoto H, Mikoshiba K (1995) The reeler gene-associated antigen on Cajal-Retzius neurons is a crucial molecule for laminar organization of cortical neurons. Neuron 14:899-912.

## References

Pagonopoulou O, Angelatou F, Kostopoulos G (1993) Effect of pentylentetrazol-induced seizures on $A_1$ adenosine receptor regional density in the mouse brain: a quantitative autoradiographic study. Neuroscience 56:711-716.

Pinheiro P, Mulle C (2006) Kainate receptors. Cell Tissue Res. 326:457-482.

Prestwich SA, Forda SR, Dolphin AC (1987) Adenosine antagonists increase spontaneous and evoked transmitter release from neuronal cells in culture. Brain Res 405:130-139.

Pritchett DB, Luddens H, Seeburg PH (1989) Type I and type II $GABA_A$-benzodiazepine receptors produced in transfected cells. Science 245:1389-1392.

Qiu S, Zhao LF, Korwek KM, Weeber EJ (2006) Differential reelin-induced enhancement of NMDA and AMPA receptor activity in the adult hippocampus. J Neurosci 26:12943-12955.

Rehavi M, Skolnick P, Paul SM (1982) Effects of tetrazole derivatives on [$^3$H]diazepam binding in vitro: correlation with convulsant potency. Eur J Pharmacol 78:353-356.

Rice DS, Curran T (2001) Role of the reelin signaling pathway in central nervous system development. Annu Rev Neurosci 24:1005-1039.

Rogers JT, Weeber EJ (2008) Reelin and apoE actions on signal transduction, synaptic function and memory formation. Neuron Glia Biol 4:259-270.

Ronzio RA, Rowe WB, Meister A (1969) Studies on the mechanism of inhibition of glutamine synthetase by methionine sulfoximine. Biochemistry 8:1066-1075.

Rothstein JD, Tabakoff B (1982) Effects of the convulsant methionine sulfoximine on striatal dopamine metabolism. J Neurochem 39:452-457.

Rothstein JD, Tabakoff B (1984) Alteration of striatal glutamate release after glutamine synthetase inhibition. J Neurochem 43:1438-46.

References

Rothstein JD, Tabakoff B (1985) Glial and neuronal glutamate transport following glutamine synthetase inhibition. Biochem Pharmacol 34:73-79.

Rouvroit de CL, Goffinet AM (1998) The Reeler mouse as a model of brain development. Adv Anat Embryol Cell Biol 150:1-106.

Rudolph U, Crestani F, Benke D, Brunig I, Benson JA, Fritschy JM, Martin JR, Bluethmann H, Mohler H (1999) Benzodiazepine actions mediated by specific gamma-aminobutyric acid(A) receptor subtypes. Nature 401:796-800.

Ruiz A, Sachidhanandam S, Utvik JK, Coussen F, Mulle C (2005) Distinct subunits in heteromeric kainate receptors mediate ionotropic and metabotropic function at hippocampal mossy fiber synapses. J Neurosci 25:11710-11718.

Scharfman HE, Gray WP (2007) Relevance of seizure-induced neurogenesis in animal models of epilepsy to the etiology of temporal lobe epilepsy. Epilepsia 48 Suppl 2:33-41.

Schmidt D, Löscher W (2009) New developments in antiepileptic drug resistance: an integrative view. Epilepsy Curr. 9:47-52.

Stone TW, Ceruti S, Abbracchio MP. (2009) Adenosine receptors and neurological disease: neuroprotection and neurodegeneration. Handb Exp Pharmacol 193:535-87.

Stransky Z (1969) Time course of rat brain GABA levels following methionine sulphoximine treatment. Nature 224:612-613.

Squires RF, Saederup E, Crawley JN, Skolnick P, Paul SM (1984) Convulsant potencies of tetrazoles are highly correlated with actions on GABA/benzodiazepine/picrotoxin receptor complexes in brain. Life Sci 35:1439-1444.

Tchekalarova J, Sotiriou E, Georgiev V, Kostopoulos G, Angelatou F (2005) Up-regulation of adenosine $A_1$ receptor binding in pentylenetetrazol kindling in mice: effects of angiotensin IV. Brain Res 1032:94-103.

References

Terashima T, Inoue K, Inoue Y, Mikoshiba K, Tsukada Y (1983) Distribution and morphology of corticospinal tract neurons in reeler mouse cortex by the retrograde HRP method. J Comp Neurol 218:314-326.

Tueting P, Costa E, Dwivedi Y, Guidotti A, Impagnatiello F, Manev R, Pesold C (1999) The phenotypic characteristics of heterozygous reeler mouse. Neuroreport 10:1329-1334.

Vincent P, Mulle C (2009) Kainate receptors in epilepsy and excitotoxicity. Neuroscience 58:309-323

Volk HA, Arabadzisz D, Fritschy JM, Brandt C, Bethmann K, Löscher W (2006) Antiepileptic drug-resistant rats differ from drug-responsive rats in hippocampal neurodegeneration and GABA(A) receptor ligand binding in a model of temporal lobe epilepsy. Neurobiol Dis. 21:633-46.

Walsh LA, Li M, Zhao TJ, Chiu TH, Rosenberg HC (1999) Acute pentylenetetrazol injection reduces rat $GABA_A$ receptor mRNA levels and GABA stimulation of benzodiazepine binding with no effect on benzodiazepine binding site density. J Pharmacol Exp Ther 289:1626-1633.

Weeber EJ, Beffert U, Jones C, Christian JM, Forster E, Sweatt JD, Herz J (2002) Reelin and ApoE receptors cooperate to enhance hippocampal synaptic plasticity and learning. J Biol Chem 277:39944-39952.

Werner P, Voigt M, Keinänen K, Wisden W, Seeburg PH (1991) Cloning of a putative high-affinity kainate receptor expressed predominantly in hippocampal CA3 cells. Nature 351:742-744.

White HS (2002) Animal models of epileptogenesis. Neurology 59: 7-14.

Yamamoto T, Sakakibara S, Mikoshiba K, Terashima T (2003) Ectopic corticospinal tract and corticothalamic tract neurons in the cerebral cortex of yotari and reeler mice.

Zilles K, Amunts K (2009a) Centenary of Brodmann's map - conception and fate. Nat Rev Neurosci. 11:139-145.

Zilles K, Amunts K (2009b) Receptor mapping: architecture of the human cerebral cortex. Curr Opin Neurol. 22:331-339.

Zilles K, Schleicher A, Palomero-Gallagher N, Amunts K (2002) Quantitative analysis of cyto- and receptor architecture of the human brain. In: Brain Mapping. The Methods (Toga AW, Mazziotta JC, eds), pp 573-602. Amsterdam: Elsevier.

Zilles K, Qu MS, Kohling R, Speckmann EJ (1999) Ionotropic glutamate and GABA receptors in human epileptic neocortical tissue: quantitative in vitro receptor autoradiography. Neuroscience 94:1051-1061.

## Danksagung

*Also tragt es in die Welt, haut es mit Edding an die Wände, so lang die dicke Frau noch singt, ist die Oper nicht zu Ende.* Kettcar, Ich danke der Academy

Ich möchte meinen Betreuern danken.
Herrn Prof. Zilles, für die Möglichkeit diese Arbeit an seinem Institut durchzuführen, ferner für das mir entgegengebrachte Vertrauen und den planerischen Freiraum, die diese Arbeit erst ermöglichten. Außerdem für die stets dezidierte und konstruktive Korrektur meiner Manuskripte. Ich danke ferner Frau Prof. Rose, die sich bereitwillig für die Begutachtung dieser Arbeit zur Verfügung gestellt hat, und mich außerdem ermutigte diese in kumulativer Weise anzufertigen. Herrn Prof. Lübke bin ich zu Dank verpflichtet für seine fachliche Beratung, seine konstruktiven Beiträge zu meiner Arbeit und für die personelle technische Unterstützung.

In jeder Phase dieser Arbeit durfte ich in einem Umfang persönliche Unterstützung erfahren, der es mir unmöglich macht, hier allen Mitarbeitern und Kollegen adäquat Rechnung zu tragen. Der Versuch soll an anderer Stelle geschehen. Ich will sie dennoch namentlich nennen, mit dem Versprechen, mich stets darauf festnageln zu lassen: Sabrina Buller, Markus Cremer, Werner Hucko, Jennifer Lopez Escobar, Stefanie Klein, Stephanie Krause, Sabine Wilms.

Ich danke.

Den Jungs von der Band für drei Akkorde gegen Schreibblockade, Tobi für Unterstützung mit dem Fliwatüüt, Markus Wiebusch für den Glauben daran und das Mittel dagegen, Matt Groening für den Friedensnobelpreis. Und der Emma, denn die wohnt hier.

Ich danke meiner Familie, besonders meiner Mutter, weil sie niemals müde werden. Und ich danke meiner Freundin, Jennifer, für die Unterstützung, die niemand anders zu leisten in der Lage wäre.

*Lieber Gott, wir danken dir für gar nichts; wir haben alles selbst bezahlt.* Bart Simpson

OK, danke, langt.

# Addendum

Cremer CM, Palomero-Gallagher N, Bidmon HJ, Schleicher A, Speckmann EJ, Zilles K (2009) Pentylenetetrazole-induced seizures affect binding site densities for GABA, glutamate and adenosine receptors in the rat brain. Neuroscience 163(1):490-9.

Cremer CM, Cremer M, Escobar JL, Speckmann EJ, Zilles K (2009) Fast, quantitative in situ hybridization of rare mRNAs using 14C-standards and phosphorus imaging. J Neurosci Methods 185:56-61.

Cremer CM, Bidmon HJ, Görg B, Palomero-Gallagher N, Escobar JL, Speckmann EJ, Zilles K (2010) Inhibition of glutamate/glutamine cycle in vivo results in decreased benzodiazepine binding and differentially regulated GABAergic subunit expression in the rat brain. Epilepsia, in press [uncorrected proof].

Die VDM Verlagsservicegesellschaft sucht für wissenschaftliche Verlage abgeschlossene und herausragende

# Dissertationen, Habilitationen, Diplomarbeiten, Master Theses, Magisterarbeiten usw.

für die kostenlose Publikation als Fachbuch.

Sie verfügen über eine Arbeit, die hohen inhaltlichen und formalen Ansprüchen genügt, und haben Interesse an einer honorarvergüteten Publikation?

Dann senden Sie bitte erste Informationen über sich und Ihre Arbeit per Email an *info@vdm-vsg.de*.

**Sie erhalten kurzfristig unser Feedback!**

VDM Verlagsservicegesellschaft mbH
Dudweiler Landstr. 99　　　　　　　Telefon　+49 681 3720 174
D - 66123 Saarbrücken　　　　　　　Fax　　　+49 681 3720 1749
**www.vdm-vsg.de**

Die VDM Verlagsservicegesellschaft mbH vertritt

Printed by Books on Demand GmbH, Norderstedt / Germany